THE NATURE OF VIOLENT STORMS

LOUIS J. BATTAN was born in 1923 in New York City. After having entered the City College of New York, he enlisted in the U. S. Army Air Force in World War II and later became a weather officer. With a group of selected Air Force lieutenants, he attended radar and electronics classes at Harvard and M.I.T. and was trained in the use of radar equipment for making special meteorological observations. Following the war, he returned to his studies. In 1946, he received his B.S. from New York University and joined the U. S. Weather Bureau where he worked for three years on the Thunderstorm Project, a research program. Graduate work at the University of Chicago resulted in an M.S. degree in 1949 and a Ph.D. in 1953. He remained there after graduation and continued his work on the physics of clouds and precipitation.

In 1958, Dr. Battan joined the staff of The University of Arizona in Tucson as a Professor of Meteorology and the Associate Director of the Institute of Atmospheric Physics. Since then he has continued to do research in the various areas of physical meteorology, including the effects of cloud seeding. Recently elected to the Council of the American Meteorological Society, Professor Battan has served on the Committee on Radar Meteorology, is currently chairman of the Committee on Severe Storms and is an associate editor of the *Journal of Meteorology*. He is the author of *Radar Meteorology* (University of Chicago Press, 1959), the first textbook to deal with this field, and many articles for scientific journals. Dr. Battan and his wife and two children live in Tucson.

THE NATURE
OF
VIOLENT STORMS

by Louis J. Battan

GREENWOOD PRESS, PUBLISHERS
WESTPORT, CONNECTICUT

Library of Congress Cataloging in Publication Data

Battan, Louis J
    The nature of violent storms.

    Reprint of the ed. published by Anchor Books, Garden
City, N.Y., in Science study series.
    Bibliography: p.
    Includes index.
    1. Storms. I. Title.
[QC941.B25 1981]   551.5'5   80-24986
ISBN 0-313-22582-6 (lib. bdg.)

ILLUSTRATIONS BY ROBERT E. RIDLEY
TYPOGRAPHY BY SUSAN SIEN

Reprinted with the permission of Louis J. Battan.

Reprinted in 1981 by Greenwood Press
A division of Congressional Information Service, Inc.
88 Post Road West, Westport, Connecticut 06881

Printed in the United States of America

10 9 8 7 6 5 4 3 2 1

THE SCIENCE STUDY SERIES

The Science Study Series offers to students and to
the general public the writing of distinguished
authors on the most stirring and fundamental top-
ics of physics, from the smallest known particles
to the whole universe. Some of the books tell of
the role of physics in the world of man, his technology and civilization. Others are biographical in
nature, telling the fascinating stories of the great
discoverers and their discoveries. All the authors
have been selected both for expertness in the fields
they discuss and for ability to communicate their
special knowledge and their own views in an interesting way. The primary purpose of these books
is to provide a survey of physics within the grasp
of the young student or the layman. Many of the
books, it is hoped, will encourage the reader
to make his own investigations of natural phenomena.

These books are published as part of a fresh
approach to the teaching and study of physics. At
the Massachusetts Institute of Technology during

1956 a group of physicists, high school teachers, journalists, apparatus designers, film producers, and other specialists organized the Physical Science Study Committee, now operating as a part of Educational Services Incorporated, Watertown, Massachusetts. They pooled their knowledge and experience toward the design and creation of aids to the learning of physics. Initially their effort was supported by the National Science Foundation, which has continued to aid the program. The Ford Foundation, the Fund for the Advancement of Education, and the Alfred P. Sloan Foundation have also given support. The Committee is creating a textbook, an extensive film series, a laboratory guide, especially designed apparatus, and a teacher's source book for a new integrated secondary school physics program which is undergoing continuous evaluation with secondary school teachers.

The Series is guided by a Board of Editors consisting of Paul F. Brandwein, the Conservation Foundation and Harcourt, Brace and Company; John H. Durston, Educational Services Incorporated; Francis L. Friedman, Massachusetts Institute of Technology; Samuel A. Goudsmit, Brookhaven National Laboratory; Bruce F. Kingsbury, Educational Services Incorporated; Philippe Le-Corbeiller, Harvard University; and Gerard Piel, *Scientific American.*

PREFACE

In the course of a few years everyone witnesses a variety of exciting weather events. To some the most striking weather story is the awesome thunderstorm with vivid lightning, wind, and rain. To others nature brings violence in the form of a tornado or a hurricane. It is reasonable to wonder why and how these storms form and how they grow and die. In this short book, I have attempted to describe various types of weather disturbances ranging from small white clouds of fair weather to hurricanes churning their destructive ways toward a coast line.

In going through this book, the reader will probably be struck by the frequent use of such phrases as "a satisfactory explanation for this is lacking" or "we still do not fully understand that." Why are there so many unsolved problems?

First of all, meteorology is a relatively young science. Most of the progress has been achieved in the last fifty or sixty years. The number of scientists who have devoted their full efforts to unraveling the many mysteries of the atmosphere has been small. Even today the demand for skilled atmospheric scientists greatly exceeds the supply. Coupled with the insufficiency of researchers has been the lack of adequate financial support for research. Many important problems have received inadequate attention because there have not existed the funds for special airplanes, radar equipment, etc. Fortunately, in the last few years, the picture has begun to brighten somewhat. More young people are choosing meteorology as a career, and more funds are becoming available.

A second major factor retarding progress in meteorology can be laid on the doorstep of the atmosphere itself. Regardless of the types of experiments a scientist may perform in his laboratory, regardless of the mathematical calculations he may make with his electronic computer, in the end he must go out and make measurements in the atmosphere. At this stage of the investigation real difficulties frequently arise. One obvious problem is that of finding the type of weather disturbance he wants to study at the right time and in the right place. Let us say he is engaged in detailed studies of hurricanes. This would require specially instrumented airplanes as well as ground-located equipment. What frustration would face him if hurricanes failed to form. Many times situations such as this actually occur.

How would you go about making measurements in a tornado knowing that one would probably form in the state of Kansas, but that it would be only 300 feet across and would last for two minutes? Then, bearing in mind that winds of several hundreds of miles per hour would be present in the tornado, what types of instruments would you use? These are the kinds of problems meteorologists are facing. The atmosphere will not stand still and be measured. This means, of course, that for the solution of some problems the scientist must be patient, but more important, must be imaginative in the conception of his attack. Over the last couple of decades meteorologists have demonstrated increased amounts of imagination and daring, and the answers are coming at a more rapid rate.

The author is grateful to Mr. Morgan Monroe of The University of Arizona for encouraging him to write this book and to Mr. John H. Durston for his excellent job of editing.

LOUIS J. BATTAN

CONTENTS

THE SCIENCE STUDY SERIES   9

PREFACE   11

1 INTRODUCTION   19

2 THERMALS AND CONVECTIVE CLOUDS   25
Convection in Clear Air—Convection in Moist Air—Thermals in Clear Air—Thermals and Small Clouds

3 AIR MOTIONS AND THE FORCES THAT PRODUCE THEM   39
Pressure Force—Centrifugal Force—Coriolis Force—Forces and the Wind—Vertical Air Motions

4 THUNDERSTORMS   53
Columnar Theory of a Thunderstorm—Bubble Theory of a Thunderstorm—Rain and Hail from Thunderstorms—Lightning—Gustiness at the Surface—Lines of Thunderstorms–Squall Lines

5   TORNADOES   75

Description of a Tornado—Why Tornadoes Are So Destructive—Some Unusual Tornado Stories—The Formation of Tornadoes—Waterspouts—Tornado Detection and Protective Measures

6   HURRICANES   99

What Is a Hurricane?—Formation and Dissipation of Hurricanes—Hurricane Eye—Rainfall Patterns in Hurricanes—Movement of Hurricanes—Hurricane Forecasting—Ocean Waves Produced by Hurricanes—Hurricane Damage and Safety Measures

7   CYCLONES   129

Difference Between Cyclones, Hurricanes, and Tornadoes—Formation of Cyclones—Regions of Cyclone Formation and Cyclone Movements

8   CONCLUSION   143

ADDITIONAL READING MATERIAL   147

INDEX   151

THE NATURE OF VIOLENT STORMS

CHAPTER 1

INTRODUCTION

Meteorology is one branch of science in which almost everyone may participate. People who work out-of-doors soon develop firm ideas on weather forecasting. Certain cloud types and wind directions mean it will rain; others mean it will not rain. Even the old-fashioned housewife who hung her clothes on the line in the back yard was prepared to match her skills with the Weather Bureau. Many of the unwritten rules of the laymen are correct, of course. The extreme conditions, perfectly clear or very rainy, are frequently quite easy to predict if the time period is short. The real problem of the weather forecaster is to be able to predict all types of weather and to do it accurately one or more days in advance. When forecasts go sour, the reasons can almost always be found in a lack of understanding of the physical processes involved, not only by the forecasters but by meteorologists in general. The unknowns in meteorology probably outnumber the knowns. Furthermore, every question answered usually produces many more questions of equal or greater interest and importance.

Among the fascinating problems confronting meteorologists are those dealing with thunderstorms, tornadoes, and other vortices in the atmosphere. The winds in such storms may vary from gusts of low speeds to whirls of air spinning at several hundreds of miles per hour. Great losses of life and property occur every year whenever violent vortices strike unsuspecting communities. Scientists have constantly sought better descriptions of storm structures and explanations for their formation. After decades of effort a great deal has been learned, not nearly enough, of course, but at least there is enough known to make useful forecasts and intelligently plan new research.

Intense vortices in the atmosphere can be taken as signs that the atmosphere is unstable and has high moisture contents. These ingredients are necessary in order to supply the huge quantities of energy needed to start and maintain the circulating air. It is easy to fail to appreciate the total power of a storm because one experiences only a small part at any one time. However, estimates of the energy involved in various weather systems can be made fairly easily if the size of the system and the wind speeds are known. Table 1 gives estimates of the *kinetic energy* of the weather systems to be discussed in the remaining chapters of this book. The kinetic energy, for those not familiar with the term, is that energy represented by the motion of the air. In order to stop the air altogether, one would have to expend the same amount of energy. It should be noted that the numbers apply to an average storm. In any particular case the amount of energy may be somewhat larger or smaller than the amount shown in the table.

Table 1

Order of magnitude estimates of the kinetic energy of various wind systems and of atomic and hydrogen bombs expressed in kilowatt-hours.

SYSTEM	KINETIC ENERGY*
Gust	Less than 1
Dust Devil	10
Tornado	10^4
Thunderstorm	10^6
Hurricane	10^{10}
Cyclone	10^{11}
Nagasaki Atom Bomb	10^7
Hydrogen Bomb	10^{10}

* The superscript refers to the number of zeros after the one. For example, 10^4 means 10,000 and 10^6 means 1,000,000.

By any scale of measure it is obvious that weather systems are powerful. This becomes even more striking when one realizes that the amount of energy input required to develop these kinetic energies is 10 to 100 times greater than the tabulated values. The major portion of the energy in atmospheric systems is expended in overcoming the effects of friction and in heating the air inside and outside the systems. It can readily be seen that the input energy to an average hurricane may be equivalent to more than 10,000 atomic bombs of the kind that destroyed Nagasaki.

The reason storms such as cyclones and hurricanes do not do more damage is that the energy is dis ibuted over a large region. In a tornado, on the other hand, it may be concentrated over an area 100 yards wide. Then explosive effects occur.

The numbers in the table also clearly indicate that man will find it difficult to modify the weather by competing with nature on the energy field-of-battle. If one were to try to produce cyclones by supplying the necessary energy, the efforts would be in vain. One could do it with hydrogen bombs but the radioactivity would make the consequences disastrous. If we are to succeed in attempts to modify large-scale weather systems, it will first be necessary to find unstable regions which can be influenced by small energy changes. We must find the equivalent of the delicately balanced boulder, which is too heavy to carry, but which will roll down the mountain if given a small push. This area of meteorology is a fascinating one and permits much speculation, but it is outside the scope of this book.

There are various sources of energy for atmospheric vortices. Heat contained in the air and earth's surface and the sinking of heavier air when it moves over lighter air are important factors. But the major contribution comes from heat released when water vapor condenses to form clouds. R. R. Braham, Jr., at the University of Chicago, has shown that in a single thunderstorm about 3 miles in diameter, there may be 500,000 tons of condensed water in the form of water droplets and ice crystals. In the course of producing these particles, there would have been released about 3×10^{14} calories. This is equal to about 10^8 kw-hrs or about 100 times greater than the value of the estimated kinetic energy given in Table 1.

Tornadoes are associated with thunderstorms

and apparently derive their energy from them. In hurricanes and cyclones there are widespread areas of cloudiness and rainfall. They indicate the release of enormous quantities of heat of condensation.

When the humidity of the air is very low, relatively small quantities of energy are available for conversion into kinetic energy. Nevertheless, vortices still develop. They may be seen as dust devils over the arid regions of the world. These whirls sometimes become large enough to blow down wooden shacks or unhinge screen doors, but they fall far short of a tornado.

The important prerequisite for the formation of a dust devil is a high temperature at the earth's surface. When this condition is combined with an air flow which has a certain amount of curvature, the whirls of the desert are a common sight. When the earth is very hot, the overlying air always is affected. Dust devils are markers of unusual activity because dust is carried upward and can be seen. If a small airplane is flown within several thousand feet of the hot terrain, the effects of the heating become dramatically evident as turbulence, even when no dust is present. Air in the lowest layers heats up, becomes lighter and rises. When an airplane passes through a parcel of ascending air it is carried upward. If the air ascends in a certain region, you usually can expect air to be descending in a nearby region. As a result, an airplane constantly experiences upward and downward accelerations. In this situation the air has a very gusty character and is said to be turbulent.

Quite evidently air motions in the atmosphere cover a wide range of sizes and intensities from weak gusts amounting to a few feet per minute to extreme updrafts with speeds of several miles per minute. The winds in the atmosphere can vary from a mile per hour or less around a large high-pressure area to several hundred miles per hour around a circle having a diameter of several hundred yards. The duration of the unusual winds can vary from gusts which last only a few seconds to vortices which last for days. Dimensions of the various phenomena also have a large range, extending from gusts of a few feet or so to hurricanes and cyclones more than 1000 miles in diameter.

In the discussion to follow various scales of motion will be examined in an effort to show how they form, grow, and die. The weather systems will be considered in order of increasing sizes and, when possible, the interactions will be pointed out.

Many links between observation and explanation have not been found yet. As a matter of fact, often the observations needed to adequately describe the weather system are still missing. These shortcomings will be mentioned at the appropriate places. It will be seen that frequently the absence of vital observations is forgivable. For example, it has not been possible to measure the extreme speeds believed to be associated with tornadoes because the passage of the storm over a weather station destroys not only the instruments but the structures supporting them as well. However, new instrument developments are now overcoming difficulties such as this one. There is every reason for believing that suitable observations will be collected, and that better descriptions than those now available will be possible in the near future. Tiros I's magnificent pictures showed us one way.

CHAPTER 2

THERMALS AND CONVECTIVE CLOUDS

If you take a vessel with a thin layer of fluid and heat the bottom, some interesting things begin to happen. At first the added heat is carried upward by molecular action. There is little motion of the fluid. At the same time the temperature difference between the top and bottom of the fluid increases. When the difference reaches a certain critical value, the fluid begins moving upward and downward in a systematic pattern. You can think of the fluid surface as consisting of a system of equal hexagons with ascending fluid in the center and descending fluid at the edges (Fig. 1). Each of the hexagonal systems is called a *cell,* and the regular pattern of cells is said to be evidence of existence of Benard convection.

The term *convection* is used to connote the transfer of heat or some other property by means of vertical motion. Where horizontal motion is involved meteorologists use the term *advection.*

The existence of convection patterns of the Benard type can easily be demonstrated in the laboratory or in your kitchen. As already noted, the important prerequisite is a relatively large temperature difference between the upper and lower layers of the fluid. If you divide the difference by the depth of the fluid, you obtain a quantity known as the *lapse rate*. The term received this name for the obvious reason that it measures the lapse or decrease of temperature for a unit rise of vertical distance or altitude. When the temperature decreases with altitude, the lapse rate is a positive quantity. When the reverse is true, it is negative.

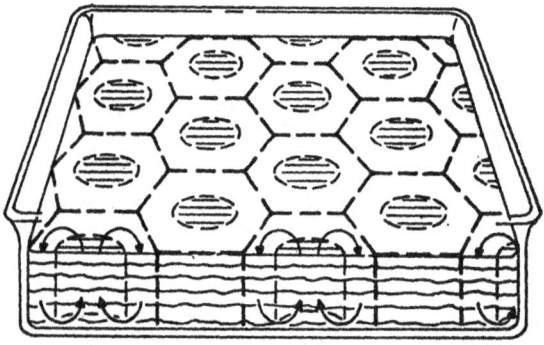

FIG. 1.   *Cellular convection, the spread of heat through a fluid, occurs in this hexagonal pattern. The fluid rises in the center of each hexagon and sinks at the edges.*

In order for Benard cells to form, it is necessary that the lapse rate have a particular positive value which depends on the properties of the fluid. When the critical value is approached, the fluid becomes unstable and convection begins. In most experiments instability is brought about by heating the bottom of the fluid. Similar results can be obtained by cooling the upper level. Convection cells frequently can be seen in a hot cup of coffee. If cold milk is carefully poured in, the milk initially settles at the bottom of the cup. Because of evaporation of the upper layer of the coffee, instability will set in, and a pattern of convection becomes evident. Small regions of mixed coffee and milk will be seen. For various reasons the patterns are irregular, but distinct areas of rising motions can be seen.

If the fluid whose lapse rate is unstable is moving horizontally, the pattern of convection is considerably different from that of our coffee and milk. When the fluid velocity increases with height, a series of rolls oriented in the direction of the flow (Fig. 2) replaces the regularly spaced cells. The fluid rises along particular lines and descends in the adjacent areas.

The precise conditions for the formation of convection in a thermally unstable fluid have been shown mathematically as well as experimentally. As the lapse rate gets larger and larger, the pattern of convection can become more and more complicated as various modes of convection become superimposed. In general, however, the simple forms already discussed predominate.

Numerous attempts have been made to extend the results of the laboratory and theoretical studies of convection to the atmosphere. In principle you need a thin layer of fluid with a rigid lower boundary and either a rigid or free upper boundary. For certain types of problems the earth can be so considered. Air can be treated as a fluid, and if proper consideration is given to its properties, it obeys the laws of fluid motion. Frequently the atmosphere contains layers of air having different properties. For example, over the tropical oceans a stable layer is usually found at 6000 to 7000 feet above the ocean. This layer acts, in some ways, like a free surface on top of the air below it. Another feature required for convection is thermal instability. Over the tropical oceans you would look for instability and convective patterns in the shallow layer below the stable level.

Convection in Clear Air

On clear sunny days incoming solar radiation heats up the ground. The transfer of heat from the ground causes a warming of the lower layers of the adjacent air. The lower layers also are warmed up because of direct adsorption of solar radiation by the water vapor. Through continual warming by these processes instability may occur.

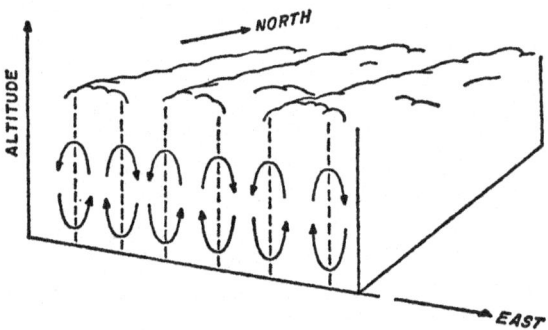

FIG. 2. *Lines of convection replace the hexagonal cells in a horizontally moving fluid with an unstable lapse rate. The curved arrows show the pattern of fluid rise and fall along particular lines.*

It is well at this point to examine in a little more detail what is meant by instability. In general, the term means that if a parcel of air is displaced from a position of rest, it will continue moving away at an accelerating rate. On the other hand, in a stable situation, a parcel displaced from a certain position and then released will return to its original position. As an illustration, a layer of oil resting above a layer of water is very stable.

In the context of this chapter we are considering vertical motions, and instability will be said to exist when a parcel of air displaced upward continues its upward motion at an accelerated pace. This will happen only if the parcel is subjected to an upward force. Suppose, for simplicity, we let the parcel be a balloon. It will accelerate upward only if it is lighter than the air surrounding it. This can be accomplished in several ways: (1) the parcel must be warmer or (2) it should contain more water vapor. The second condition may be somewhat startling to some, but it follows simply from the fact that the molecular weight of water vapor is less than that of dry air. The former is about 18; the latter is 29. If the pressure and temperature are constant, a given volume of dry air will be 29/18 heavier than the same volume of water vapor. In general though, this factor is of less importance in determining stability than temperature differences. For example, a relative humidity difference of 50 per cent at a temperature of 32° F and an altitude of 10,000 feet corresponds to the same effect as a temperature difference of only 1° F.

If the atmosphere is to be unstable, it is not

sufficient merely to have a temperature that decreases with altitude. If this were so, the air would almost always be unstable because the temperature almost always decreases upward. For instability the air parcel ascending through air having a positive lapse rate must always be less dense than the environment, that is, warmer than the air surrounding it. In a liquid, expansion of a parcel of fluid can usually be ignored, but with air the reverse is true. When air ascends, it moves into regions of continuously decreasing pressure. As a result it expands and cools. If there is no addition or loss of heat and no condensation, the parcel temperature decreases at a rate of about 5.5° F per 1000 feet of ascent. This is known as the *adiabatic lapse rate*. Most air motions in the clear free atmosphere follow the adiabatic law.

We now have enough information to specify those conditions that may lead to instability in the atmosphere. We already have said that a rising parcel cools at a rate of 5.5° F per 1000 feet. In order for it to remain warmer than its environment, the latter must have a lapse rate exceeding the dry adiabatic rate. In this circumstance strong and prolonged vertical motion is assured.

If a parcel of air near the earth's surface becomes very warm, strong vertical motions may occur in a stable atmosphere. The parcel ascent will continue until its temperature reaches that of the environment. This type of situation can be observed when huge amounts of heat are released. Dramatic examples are seen when tall clouds form over forest fires, oil refinery fires, or the explosion of nuclear weapons.

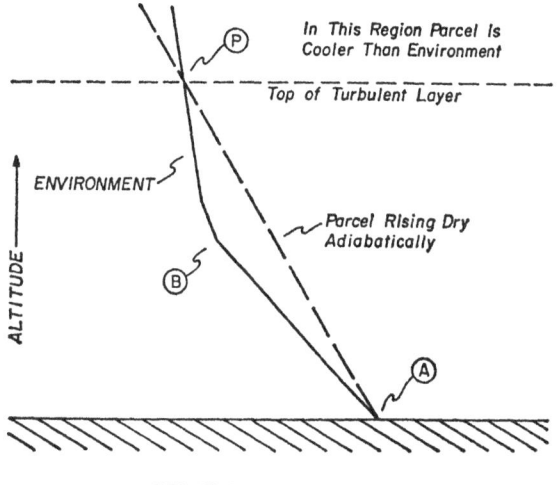

In This Region Parcel Is
Cooler Than Environment

Top of Turbulent Layer

ENVIRONMENT

Parcel Rising Dry
Adiabatically

ALTITUDE

TEMPERATURE ⟶

FIG. 3.   *A rising parcel of air warmer than the surrounding atmosphere creates turbulence that torments queasy plane passengers. In this diagram the solid line shows how the temperature of the stable environment varies with altitude. The broken line charts the dropping temperature of the dry parcel as it rises. At P the temperature of parcel and environment is the same. The air is unstable from altitude A to altitude B.*

A common feature of the atmosphere over the deserts is instability of the lapse rate in the layer of air near the earth's surface (Fig. 3). Here the vertical motion will be restricted to the layer below the level designated as the top of the turbulent layer. With an airplane we can identify such a layer quite easily. The two very noticeable manifestations are the turbulence and haze.

The small-scale turbulence that causes an airplane to bounce and shake is an indication of vertical air motion. The same air motion carries dust and smoke from terrestrial sources up to higher layers of the atmosphere. The upper level of the dust and turbulence indicates the upper limit of the convection initiated close to the ground.

Convection in Moist Air

A rising parcel of air, we have seen, is cooled by expansion. The cooling results in an increase of the relative humidity of the air. If there is sufficient moisture in the parcel and the ascent is sufficiently great, the humidity will approach 100 per cent, at which time condensation begins. At this point a cloud becomes visible. It signals a release of the heat of condensation which makes the parcel less dense. For every gram of water vapor condensed about 600 calories of heat are added to the rising parcel. This effect tends to compensate partially for the cooling caused by expansion. If the parcel continues ascending, its temperature decrease will now be about 3.3° F per 1000 feet as contrasted with 5.5° F per 1000 feet. For obvious reasons the new lapse rate is called the *moist adiabatic rate*.

Once condensation has begun, conditions for instability become less restrictive. An environment with a temperature lapse rate greater than the moist adiabatic rate is unstable (Fig. 4). Except in shallow layers near the earth's surface, lapse rates exceeding the dry adiabatic one are seldom found. On the other hand, it is common to find them greater than the moist adiabatic rate.

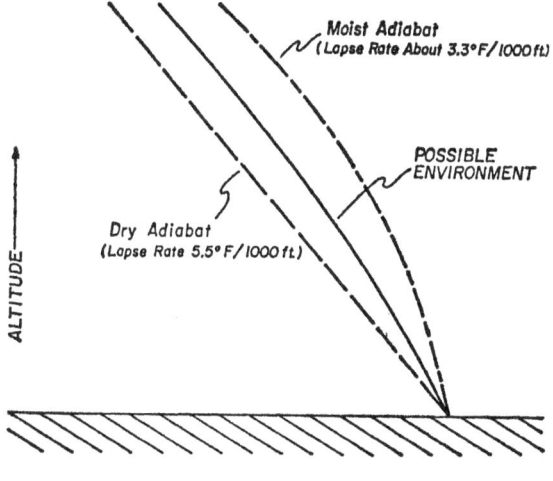

FIG. 4. *If the atmosphere had a lapse rate such as the one labeled "possible environment" it would be unstable if a saturated parcel of air were rising: the parcel would follow the moist adiabat and be warmer than the environment. On the other hand, a parcel of dry air would follow the dry adiabat and be cooler than the environment. In this case the environment would be considered to be stable.*

Thermals in Clear Air

A thermal is a rising body of air which may carry various objects upward at rates ranging from less than an inch per second to more than 100 feet per second. The largest updraft values are found in thunderstorms, but even in clear air upward speeds of tens of feet per second have been noted. In this section we shall consider thermals in cloudless air. The more violent thermals associated with clouds, thunderstorms, and tornadoes will be discussed later.

As already noted, convection cells in the atmosphere may form when an unstable fluid is heated at the bottom or cooled at the top. In general, in the atmosphere the first process is the most important, and the air for the most part is heated in two ways: (1) by passing over ground warmed by the sun's rays and (2) by passing over warmer water.

The existence of convection patterns formed as cool air is warmed from an underlying body of water was clearly demonstrated about twenty years ago by A. H. Woodcock, of the Woods Hole Oceanographic Institution. His observations were brilliantly simple. He was aware that during the winter months the air moving off the northeast coast of the United States usually is colder than the water, and he suspected the presence of convection cells. His observations were ingenious. He watched and recorded the soaring routines of herring gulls. When the winds were light, the birds soared in circular patterns which delineated the thermals and showed them to have a tilted columnar structure. When the winds were moderate, the convective motions, as indicated by the soaring of the sea gulls, were "vertical sheets extending indefinitely up and down parallel with the winds." When the air was warmer than the ocean water, the gulls did not soar. The results of these observations clearly are in agreement, at least qualitatively, with the laboratory experiments described earlier.

In subsequent work Woodcock and his colleague, J. Wyman, conducted experiments with smoke trails released from airplanes and from the surface over the tropical oceans. They again noted distinct cellular convection patterns whose dimensions were roughly in agreement with those we would expect with Benard convection.

Breeding grounds for thermals of the type emanating from surfaces warmed by solar radiation are the arid regions of the southwestern United States as well as those in other parts of the world. In the early summer months the temperature of the earth's surface may exceed 150° F, and in the lowest layers of air the lapse rates may be many times greater than the dry adiabatic rate. Conditions here are ideal for deep thermals to form. When a parcel of air gets a slight upward motion, it will accelerate rapidly because of the strong instability. An airplane flight over these very dry regions quite obviously shows the presence of thermals as the airplane is jolted upward and downward. In general the thermal diameters are small; the planes seldom gain or lose much altitude, but you have the feeling of flying over the surface of a washboard.

A common feature of the hot dry desert is the dust devil. This phenomenon is a visible manifestation of the presence of a particularly intense thermal. If air with a slight rotational motion rises because of buoyancy effects, other air rushes in to replace it. The converging air begins rotating more rapidly as it moves into the center of the whirl. For similar reasons an ice skater spins faster when he brings his arms close to his body. Properly speaking, the angular momentum of the air is conserved, and as the radius of the circle of rotation is reduced, the velocity of motion is increased. You will hear more of this in the next chapter.

Dust devils can be seen because they carry sand particles and other debris to high altitudes. They sometimes extend to over 5000 feet.

Sea gulls as well as other soaring birds have pointed out the existence of regions where thermals occur with high frequencies. Since the advent of manned gliders and soaring planes, a great deal has been learned about locations favorable for thermal formation. In addition to those areas already mentioned, convective currents usually occur over hills and mountains.

At one time it was believed that the mountain's major part was that of acting as a barrier and forcing air upward. In this sense it would be producing forced convection rather than thermal convection. It is now realized, however, that a mountain also plays an important role in setting off thermal convection. When the sun's rays fall on a mountain, the soil and rocks are heated and in turn heat the air by an amount which is higher than the radiative heating of the air at the same altitude over the adjacent valleys. As a result the air over the mountains becomes warmer and less dense, and convection can be started. The evidence for this point of view is mounting. One of the strongest arguments in its favor is the fact that the convective clouds which show the presence of upward motion usually do not begin to form until after the sun has been shining for a number of hours. In general the clouds start forming between 10 A.M. and noon. The absence of clouds during the cool morning hours indicates that the presence of an obstruction is not sufficient to establish thermals and convective clouds.

In summary it can be said that because of thermal instability and heat added near the surface of the earth convective currents are initiated. The magnitude of the motion can range from small, weak gusts which are barely detectable with sensitive instruments, to the whirl of sand known as the dust devil.

Thermals and Small Clouds

Some paragraphs back we pointed out that prolonged upward movement of moist air can lead to the formation of clouds. Over mountainous areas convective clouds, known as *cumulus,* are a common sight, especially when moist air covers the region. The organization of the clouds in mountainous areas is frequently determined by the topography.

Of somewhat more interest, from the point of view of understanding the convection process and of testing hypotheses derived from laboratory and theoretical studies, are the clouds over uniform terrain and large bodies of water.

Woodcock's observation of the lines of soaring gulls oriented in the direction of the wind immediately raises the question whether cumulus clouds are so oriented. The observations show that indeed they are. In fact certain meteorologists for more than twenty years have made use of the expression *cloudstreet* to designate lines of cumulus clouds. J. Kuettner, an expert meteorologist and glider pilot, has recently summarized much of the available information on the subject. He has pointed out that glider pilots have made use of the convection in cloudstreets for carrying out flights of more than 300 miles. The technique employed is to fly down the line and make use of the rising air to maintain altitude while keeping the glider nose pointed downward in order to maintain high airspeed. Examples of the orientation of cumulus clouds in long parallel lines are shown in Plate I. Similar types of patterns over vast regions of the tropical Pacific Ocean have been found by H. Riehl and J. S. Malkus.

The main features of the environment in which these cloud lines occur are a thermally unstable atmosphere and a wind whose direction varies little with altitude. Kuettner has shown that conditions for the creation of lines of cumulus clouds parallel to the wind direction are favorable when the wind speed is at a maximum in the cloud layer. This condition also favors long glider flights because the tail wind increases the ground speed of the aircraft. He has suggested that the orientation of the cloudstreets can be used to indicate the direction of the wind velocity in the region where the clouds are formed.

Photographs taken from rockets at altitudes of several hundred miles clearly show that lines of clouds are more prevalent than once believed. Satellite observations, such as those of Tiros I and Tiros II, are yielding information which will make it possible to study the variabilities of the cloud lines in both time and space.

CHAPTER 3

AIR MOTIONS AND THE FORCES THAT PRODUCE THEM

The characteristic feature of the ocean of air in which we live is its restlessness; it almost never stops moving. Occasionally the movement slows down to the point where conventional instruments fail to record it, and the air is said to be calm. But if sufficiently sensitive devices are used, even then motion is found. At the other extreme, when storms are overhead, the movement of air is evident as strong winds carry heavy objects long distances.

Why does the wind blow?

If we could put a tag on a small volume of air, we would see that its movement was three dimensional. The horizontal motion is usually much greater than the vertical motion, but the latter does exist, and, as we showed in Chapter 2, is of vital importance in the production of clouds and storms. In a later section we shall examine vertical drafts in clouds in some detail. Just now let us concentrate on the horizontal air motion which, by convention, is called the *wind*.

As is true for any physical body, air moves because of the action of various forces, in particular pressure forces and friction. In addition, curved wind flow and rotation of the earth enter the picture, and the air flow deviates from the direction it would take if pressure and friction forces alone were acting to produce straight flow on a stationary earth.

Pressure Force

The effects of the pressure force are most important and are easily measured. In the pressure

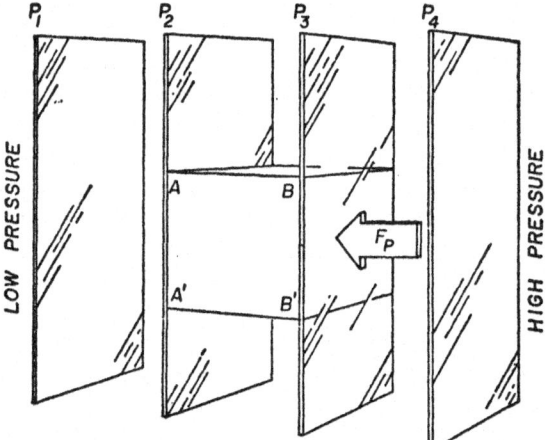

FIG. 5. *The action of pressure force is shown geometrically in this diagram.* P_1, P_2, P_3 *and* P_4 *are isobars, or lines of equal pressure; the pressure along each isobar is everywhere the same. The pressure force,* F_p, *is exerted in the direction from higher to lower pressure and is perpendicular to the isobars.*

distribution shown in Fig. 5, where P_1, P_2 etc., represent lines of equal pressure (*isobars*), the direction of the pressure force is along the line F_p. It is evident that it is from higher to lower pressure and is in the direction perpendicular to the isobars. We can obtain the magnitude of the force by considering the effects of the pressure on the unit cube A, B, B', A', one inch on the side. The force on the cube face AA' is P_2 because pressure is defined as force per unit area. The force on face BB' is P_3. The net force on the cube would be the difference $P_3 - P_2$ and would be directed as shown. The decrease of pressure over a unit distance is known as the *pressure gradient*. If we consider the force on a unit mass of air, we can write the simple equation

$$F_p = -\frac{1}{D} \times \frac{P_3 - P_2}{B - A}$$

where D is the density of the air, that is the mass per unit volume, and the ratio $\dfrac{P_3 - P_2}{B - A}$ is the pressure change per unit distance perpendicular to the isobars. The minus sign enters the equation because the force is directed from high to low pressure.

If there were no other force acting on the cube of air, it would move in the direction of F_p, and its acceleration would be calculated from the well-known formula, force equals mass times acceleration.

Note here that friction would act to slow down the air motion. In this simple case friction may be represented as a force in a direction opposite to that of the pressure force. For the time being, we shall ignore frictional forces and point out their importance later.

Centrifugal Force

Up to this point it should be clear that on a hypothetical, flat, stationary earth with straight isobars it would be an easy matter to calculate the magnitude of the winds. Unfortunately for those who have to make such calculations, the earth is not flat, it does rotate, and the isobars usually are not straight. We must accept complications.

The effects of curvature of the air flow are easily visualized. If you tie a heavy object to a string and swing it in a circle, you feel an outward pull on the string. The reason is that if the object were not tied to the string, it would move in a straight line. If the object is to continue in a circle, the string must apply an inward force. This force is called the *centripetal force*. If the string were to break, the object would continue in a straight line and would move farther away, leaving you with the end of the string at the center of the circle. Because the object does move outward when released, it is said to be under the action of a *centrifugal force*. To be precise, however, you should recognize that there is *not* a force directing the object outward.

If the air parcel on a stationary earth followed a circle, the parcel would be subject to a centrifugal force in the same way as the object at the end of the string. See Fig. 6. It can be shown that the force exerted on a unit mass of air moving

with velocity V around a circle of radius R is given
by the formula

$$F_0 = \frac{V \times V}{R}$$

Thus, the higher the velocity of the air parcel
and the smaller the radius of the circle, the greater
would be the outwardly directed centrifugal force.

Let us return to the object of mass m rotating in
a circle at the end of a string. If you let the string
wind around your finger, you will see that as the
string gets shorter, the object moves at ever-
increasing speeds. This occurs because to shorten

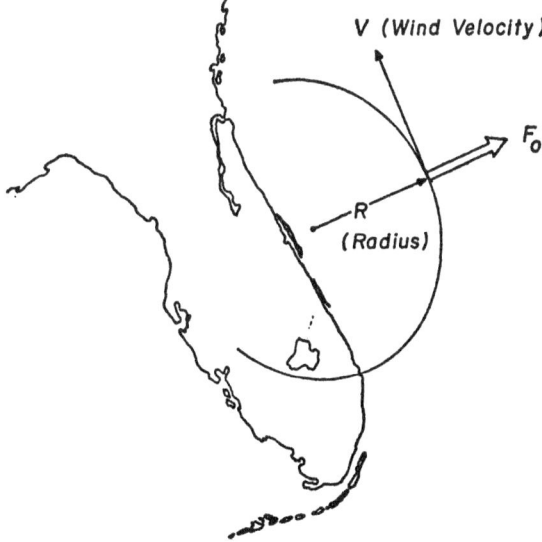

FIG. 6. *On a stationary earth an air parcel moving with the velocity of the vector* V *on a circle
would be subject to the centrifugal force shown
in this diagram at* F_0.

the string it is necessary to increase the inward directed force, that is, the centripetal force. This increase is accompanied by an increase of the kinetic energy of the moving object. For the same reasons, when an air parcel is constrained to move around a circle of decreasing radius, its kinetic energy increases. Through fairly elementary mathematics it can be shown that if the only force introduced is an increase of the centripetal force, the product of mass, velocity, and radius of the circle is *constant*. The product $m \times V \times R$ is defined as the *angular momentum*.* The principle of the conservation of this quantity plays an important part in explaining many of the vortices found in the atmosphere. It follows from this law that as a parcel of air approaches the axis of rotation, the speed of rotation increases.

Coriolis Force

An object or a parcel of air moving in a straight line over a stationary earth will continue moving in a straight line unless an outside force is exerted on it. This is not so when the earth rotates. It has been found that in the Northern Hemisphere moving air is deflected toward the right. The deflection occurs because the air moves over a rotating earth but the wind flow is measured with respect to the earth. The effects are sometimes difficult to visualize. An analogy may make the picture clearer.

Consider yourself at the center of a huge table which is turning counterclockwise at a steady rate.

* In meteorology one frequently considers the movement of a parcel of unit mass. In this case the principle of the conservation of angular momentum may be written $V \times R$ = constant.

See Fig. 7. Along one radius paint a line of crosses so that the movement of the turntable can easily be followed. When the crosses pass point *P'* shown in Fig. 7A, throw a ball toward it. If you were looking down from a fixed position above the table, you would see the ball moving in a straight line. On the other hand, if you were standing at the center of the turntable, and turning with it, and were looking down the line of crosses, the ball

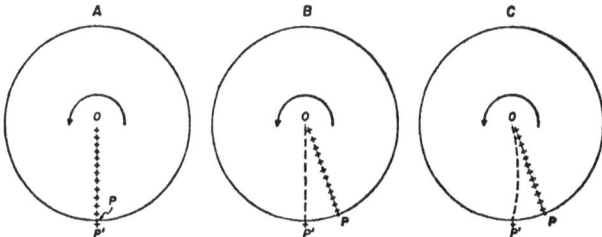

FIG. 7. *The Coriolis effect may be regarded as a "fictitious force" and may be visualized with the help of the diagram. In* (A) *an observer at O at the axis of a turntable rotating in a counterclockwise direction tosses a ball towards* P, *a point on the rim of the turntable.* P *is on a direct line from* O *to* P' *a fixed point beyond the turntable. Assume that the ball is thrown at such a speed that it would reach* P' *in the time needed for point* P *to move to the new position shown in* (B). *A fixed observer looking down on the turntable would see the ball travel along the straight dotted line and land at* P' *as shown in* (B). *On the other hand, the observer at* O *who is fixed on the table and rotating with it sees something quite different. As shown in* (C), *he sees the ball follow a curve to the right before it lands on* P'.

would appear to have followed a curved path. It would appear that the ball was deflected to the right of the direction in which it was thrown. The throw was along the series of crosses but the ball's path diverged to the right of this line by ever-increasing distances.

In a manner somewhat similar to that of the thrown ball on a turntable, air parcels on a rotating earth are deflected to the right. It should be evident that this results not because a force is applied, but rather, because the earth is moving under the air and because we measure the wind from a fixed platform mounted on the earth. The effect of the earth's rotation was first discussed by a French scientist, G. G. Coriolis, in 1835, and is called the *Coriolis effect* or the *Coriolis force*. It should again be noted that it is not a force in the sense that pressure is a force. It comes about because of the way we measure the wind. Mathematically, however, it has the properties of a force and hence can be considered one for most purposes.

The important property of the Coriolis effect is that in the Northern Hemisphere it causes the air flow to be deviated toward the right while in the Southern Hemisphere the reverse is true. The magnitude of the Coriolis force is zero at the equator and reaches a maximum value at the poles. Its magnitude at these points is equal to twice the product of the wind velocity, V, and the angular velocity of the earth, A. At intermediate latitudes, the Coriolis force can be calculated from

$$F_c = 2 \times V \times A \times sine\ l$$

where l is the latitude. See Fig. 8.

The Coriolis force plays the major role in determining the direction of air motion in many vortices in the atmosphere. This certainly is true of all the large, persistent ones. In some small ones, like dust devils, the movements of air occur over small distances and last for very short time periods, and the Coriolis effects are not sufficient to control the direction of rotation. On the other hand the circulations of hurricanes, cyclones, and anticyclones are governed by the Coriolis effect, as we shall see.

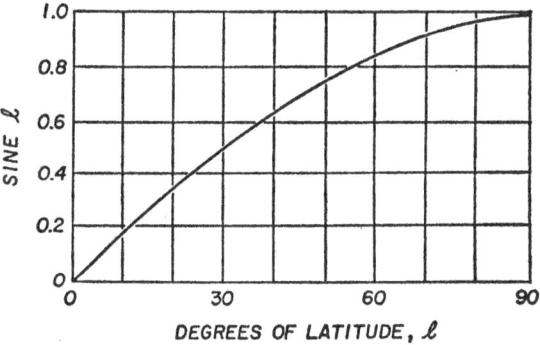

FIG. 8. *The Coriolis effect is zero at the equator and rises to a maximum value at the poles. Since the value of the effect is directly proportional to the sine of the latitude, this curve shows the variation with latitude.*

Forces and the Wind

From this brief discussion you will have an idea of how various real and apparent forces may affect the wind. The role each plays is illustrated in Fig. 9. Consider a parcel of air lying in a circular pressure field surrounding a low-pressure area. The isobars are labeled in units of millibars, the conventional system used by meteorologists (1 inch of mercury equals 33.9 millibars).

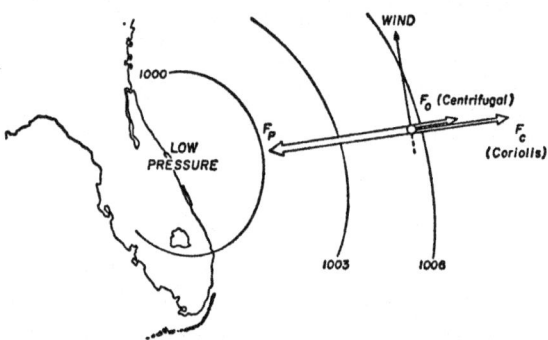

FIG. 9. *How air moves in adjustment to the pressure force and the fictitious centrifugal and Coriolis forces is seen in this diagram of a pressure pattern over the Florida coast. The arcs are isobars labeled in millibars. The stream flow resulting from the balance of pressure force and the fictitious forces follows the isobars, and in the Northern Hemisphere the low pressure always is at the left of an observer looking downwind.*

At the initial instant the parcel would begin to move in the direction of the pressure force, but as soon as it starts, the Coriolis effect comes into play and deviates the parcel to the right. If the isobars were straight, the acceleration of the air and the deflection would continue until the pressure force is just balanced by the Coriolis force and the wind is parallel to the isobars. In this circumstance the wind is said to be *geostropic*. In the Northern Hemisphere the wind direction adjusts itself so that low pressure is on the left side of an observer looking downwind.

In the situation shown in Fig. 9 the curvature of the isobars tends to cause the stream flow to have the same curvature. This leads to a centrifugal action directed outward. The air motion adjusts itself until the pressure force just balances the sum of the Coriolis and centrifugal forces. When this occurs, meteorologists refer to it as the *gradient wind*. It always will follow the isobars, and its speed will increase as the pressure force increases. As shown in the formula on page 41 of this chapter, this pressure force is proportional to the pressure gradient. The closer the isobars the higher the wind speeds.

A real force that has been neglected up to this point is friction. In general, its magnitude is small, but it sometimes leads to important consequences. As stated earlier, friction has the effect of slowing down the wind. When this happens, it causes a reduction of the Coriolis and centrifugal forces but does not affect the pressure force. As a result the wind is deflected toward lower pressure. Around a low-pressure area in the Northern Hemisphere the wind flows counterclockwise and to a large extent follows the isobars, but one always finds a small component directed inward toward the center of the vortex.

Vertical Air Motions

The vertical components of air motions usually are much smaller than the wind velocity, but since they may produce clouds and rain, they are important. Ascent or descent of air can come about in various ways. Obstacles to horizontal air flow, such as mountain ranges or the frontal surfaces we shall discuss later, are common causes of widespread upward air movement. In some weather situations the winds in the low levels of the atmosphere converge toward a particular area and cause ascending air motions. An example of this type of situation is a low-pressure area of the type discussed in the preceding section. The effects of friction cause inflow toward the center and lead to rising air. In general, the speeds of ascent in large weather systems are small, of the order of an inch per second.

The strongest updrafts and downdrafts are found in thunderstorms. They come about from large instabilities of the type we discussed in Chapter 2 and will enlarge on in the next chapter. The basic condition for ascent or descent is that the air in a cloud is lighter or heavier than the air outside the cloud. The force causing the vertical motion is known as the *buoyancy force*.

The buoyancy principle is essentially a restatement of the ancient principle of Archimedes. A body immersed in a fluid is subject to a force equal to the weight of the fluid that the body displaces. We may consider that a volume of air in a cloud has displaced an equal volume of air having the properties of the air outside the cloud. If the cloud air is lighter than the air it displaced, it would be subjected to an upward force equal to the differences of the weights.

The weight of a volume of air can be calculated if you know its density. The density in turn depends on the pressure, temperature, and water-vapor content. Since the calculation of buoyancy force involves a comparison of the properties of adjacent bodies of air at the same altitude, you can safely assume that the pressure affects cloud and clear air by equal amounts and thus may be neglected. As mentioned in Chapter 2 (page 30) increasing water-vapor contents lead to decreasing air density because the molecular weight of water vapor is about 18/29 that of dry air. An increase of water vapor has the same effect as increasing temperature. It has been found convenient to create a new term called the *virtual temperature*. This quantity represents the actual temperature plus a small amount depending on the water-vapor content of the air. The additional amount ranges from zero, when the air is perfectly dry, up to several degrees, when the humidity is high.

It can be shown that the buoyancy force is directly proportional to the difference of the virtual temperature inside and outside the cloud. A knowledge of the force permits calculations of acceleration for any given mass from the well-known formula, force equals mass times acceleration. As shown in Chapter 2, if we know the variation of temperature with altitude and know the moisture content of a rising parcel, we can estimate the difference of temperature as a function of an altitude. This information makes it possible to calculate the changes of vertical speed of the air in a cloud.

You might ask what the effects of suspended water and ice particles are on cloud buoyancy. They raise the mass of the cloud air and thereby decrease the buoyancy and can be calculated if the mass of the particles is known.

It has been proposed that when the mass of accumulated water and ice particles in an updraft becomes sufficiently large, the downward force can lead to descent of air. Once the air in a thunderstorm begins descending, evaporation of cloud and precipitation particles causes a reduction of the virtual temperature. This leads to a negative buoyancy. The cloud air becomes heavier than surrounding air, and the downward force causes acceleration of the downdraft. When cold air strikes the ground and spreads out at the bottom of a thunderstorm, we experience strong, gusty winds.

The important role played by strong vertical motions in the development of severe storms will become more evident in the following chapters.

CHAPTER 4

THUNDERSTORMS

Small cumulus clouds of the type discussed in Chapter 2 are a common sight in many parts of the world at all seasons of the year. They are frequently a characteristic part of the fair weather sky. These clouds show that there has been some instability close to the surface, generally caused by air moving over warmer land or water. Each cloud lasts only a short time, five to ten minutes, and seldom grows more than a few thousand feet tall. Further cloud growth is inhibited by the stability and dryness of the air in the middle layers of the atmosphere.

When the atmosphere is unstable to great depths and the moisture content is high, convective cloud development, once started, proceeds at a rapid rate. At times several cumulus clouds form because of a local hot spot or mechanical lifting by a hill or mountain and merge into a single convective unit. The cloud air, because of its buoyancy, continues rising. In a very unstable air mass, one where the lapse rate is large, the rising parcel of air becomes more buoyant with altitude. This results because the temperature difference between the rising parcel and the environment increases with altitude. As long as this is true, the cloud air ascends at an accelerating rate. In some cases the temperature difference continues to increase to over 30,000 feet, and the cloud air may be warmer than the environment air up to the lower layers of the stratosphere.

A parcel of cloud air, ascending at the rate of perhaps 200 feet per minute at the level of the cloud base, say 5000 feet, may attain upward speeds of 5000 feet per minute or more by the time it reaches 25,000 feet. In this manner small cumulus clouds become bigger ones, known as *cumulus congestus* and these, in turn, can develop into *cumulonimbus clouds,* better known as thunderstorms. Measurements of the extreme values of updraft speeds have been few, but pilots with long experience in flying through thunderstorms certainly can attest to the existence of powerful updrafts.

Indirect evidence has permitted some estimates of how large the updrafts may be. With radar equipment it has been possible to obtain measurements of the heights of the precipitation particles in thunderstorms. Such observations should be a slight underestimation of the heights of the visual cloud tops. It has been found that the vertical extents of thunderstorms are considerably greater than would have been expected some fifteen years ago. It is not unusual to find thunderstorms that extend to altitudes over 40,000 feet (Plate II). Some radar observations have placed the tops of some extreme thunderstorms at altitudes over 65,-000 feet.

The upper limit of thunderstorm growth is determined by the height of the stratosphere. This is so because the lower layers of the stratosphere are very stable. As a matter of fact the base of the stratosphere, known as the *tropopause,* is defined as the level at which the temperature begins to increase with height or at least decreases at a rate below 1.1° F per 1000 feet (Fig. 10). When the

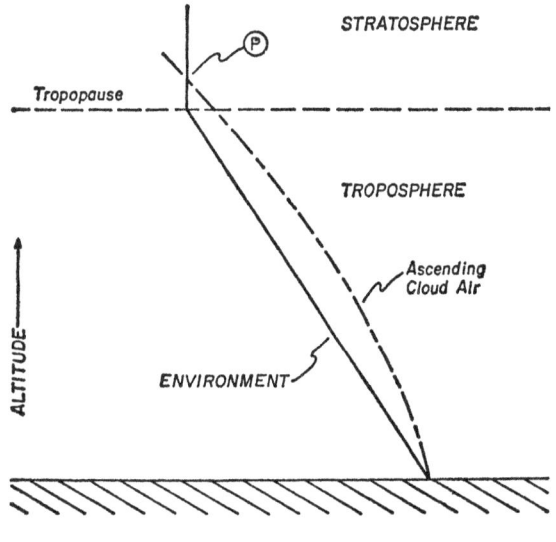

FIG. 10. *The stratosphere, which limits the height of thunderstorms, is defined as the level at which temperature begins to increase with altitude, as shown here in the environment temperature-altitude relationship (the solid line on the graph). Above the tropopause the ascending cloud is colder than the surrounding air, and its consequently increasing weight slows it down.*

rising cloud penetrates the stable layer of air, it soon finds itself in a region where the cloud is colder than the environment. Since it is now heavier than the surrounding air, there is a downward force impressed on it, and the cloud begins to slow down. Because of its momentum, it will continue upward several thousand feet even though the force is downward, but in short order it comes to a halt.

We can express the behavior of the rising cloud air in terms of an equation involving the speed of the air, and its buoyancy (see Chapter 3). J. E. McDonald has found that if reasonable values of temperature are considered, and if the cloud updraft at the base of the stratosphere is about 5000 feet per minute, the cloud should not penetrate more than about 5000 feet above the tropopause.

The height of the tropopause varies with both latitude and season of the year. During the thunderstorm season it ranges from perhaps 40,000 to 60,000 feet. Most radar observations of thunderstorm tops reveal that they are within the 5000-foot limit, but a few extend as much as two or three times as high. From these observations we must conclude that updraft speeds in the upper levels of these extreme thunderstorms could be as much as two or three times the value cited, as high as about 2 or 3 miles per minute.

Another indication of extreme updraft speeds in thunderstorms has been observations of hailstones the size of "baseballs" at 40,000 feet. Since hail will be discussed in a later section, little will be said here except to point out that a 3-inch hailstone has a fall speed of about 6000 feet per minute. Perhaps vertical motions of this magnitude do exist.

Columnar Theory of a Thunderstorm

To a casual observer, actively building convective clouds may appear to be a churning, confused mass of unrelated air motions. Detailed studies by means of specially instrumented airplanes, radar and other equipment, have shown this is not so.

A massive attack on this problem was made by the Thunderstorm Project under the leadership of H. R. Byers of the University of Chicago. This research program, carried on from 1945 to 1950, lifted much of the mystery from the awesome phenomenon known as the thunderstorm. Before this project, aircraft were vigilantly steered away from thunderstorms in fear that the turbulence, lightning, and hail might destroy the planes. One of the important results of the investigation was to establish that pilots should continue to show proper respect for the power of these storms, but that a properly designed airplane, if correctly handled, could be flown safely through thunderstorms.

One of the important contributions of the Thunderstorm Project reported by Byers and his colleague, R. R. Braham, Jr., was the concept of the *thunderstorm cell.* They concluded from a study of a large mass of observations that a thunderstorm is composed of from one to many cells, each of which passes through a fairly well-defined life cycle. In the early part of the cell history, the air motion is almost entirely upward (Fig. 11). Much of the cloud air comes from below the cloud base but some also is drawn through the sides. Throughout this so-called *cumulus stage,* the cloud builds rapidly and the updraft speeds increase. The

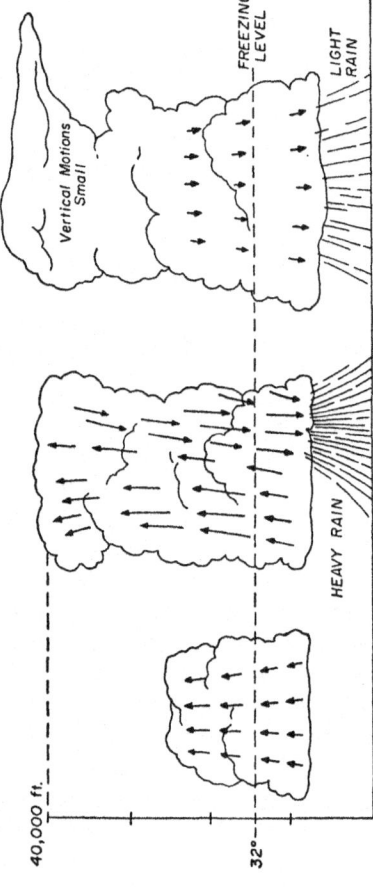

FIG. 11. *The life cycle of the thunderstorm cell is illustrated here. At the left is the cumulus stage of cloud building, in the center the beginning of the mature stage and at the right the mature stage. From H. R. Byers and R. R. Braham, Jr., The* **Thunderstorm,** *U. S. Government Printing Office.*

growth of the cloud is accompanied by the growth
of precipitation particles. When the particles become sufficiently large, their weight becomes sub-
stantial, and they exert sufficient drag to cause part
of the cloud air to start descending. This is considered the beginning of the so-called *mature stage*.
Once the downdraft has started, it speeds up rapidly. The descending air, cooled by evaporation of
precipitation, becomes heavier than the air outside
the cloud. This situation favors downward acceleration of cloud air.

During the mature stage the vertical motions,
both upward and downward, are most severe. Part
of the cloud air is rising at its maximum speed
while an increasing part of it is descending at its
maximum rate. An airplane flying through a thunderstorm at this time would experience large altitude changes, first an increase of altitude and then
a sudden decrease, or vice versa, depending on
where it entered the storm. Superimposed on the
upward and downward motions would be small-
scale gusts, which would cause the airplane to
bounce and shake.

It should be noted that the air motions can be
considered as a steady current causing altitude
changes, and an unsteady part causing the turbulence. One can think of a river as an analogy. A
boat moves smoothly down the stream without
jostling unless there are ripples and eddies to cause
it to bounce and shake. The dangerous aspects of
a thunderstorm, as far as airplanes are concerned,
are not the steady drafts but the gusts. They are
small and exert their effects for only a short time,
but their forces are large and in special cases are
capable of causing structural damage.

The mature stage of a thunderstorm is characterized by maximum rain, electrical effects, and gusts at the earth's surface.

As the region of downdraft spreads through the cloud, the energy supplied by the updraft is gradually diminished. When the entire cloud consists of descending air, the thunderstorm has reached its *dissipating* and final stage. At this time turbulence, rain, and lightning intensity have all diminished. All that remains is a large fuzzy mass of cloud which begins to evaporate rapidly.

It is conceived that each cell has a diameter of a few miles and lasts less than an hour. However, a large thunderstorm may consist of many cells, each in a different stage of development. As one cell dissipates, a new one develops, and in this way a thunderstorm may have a duration of many hours.

Bubble Theory of a Thunderstorm

A second theory of thunderstorm structure was advanced in the early 1950s, particularly through the efforts of R. S. Scorer and F. H. Ludlam. They arrived at their conclusions through careful observations of convective clouds and by means of a series of clever laboratory experiments. The experiments consisted of studies of the behavior of a heavy fluid when dropped into a lighter fluid. In some respects this is an imitation, though in reverse, of convection in the atmosphere. They found that the geometry of the spread of the fluid bore a striking resemblance to the behavior of convective clouds in the atmosphere.

The so-called *bubble theory* visualizes that a convective cloud is not a single tall column of rising air, but rather that it is composed of a series of fairly discrete "bubbles" of rising air trailing one behind the other (Fig. 12). In this model of convection the distributions of vertical motion, temperature, and quantities of cloud water are different from those pictured by the Thunderstorm Project. According to the column or jet theory, as it is sometimes called, the content of water in the liquid or ice phase, as well as the updraft speed and temperature excess of the rising air, should increase from the cloud base to near the cloud top.

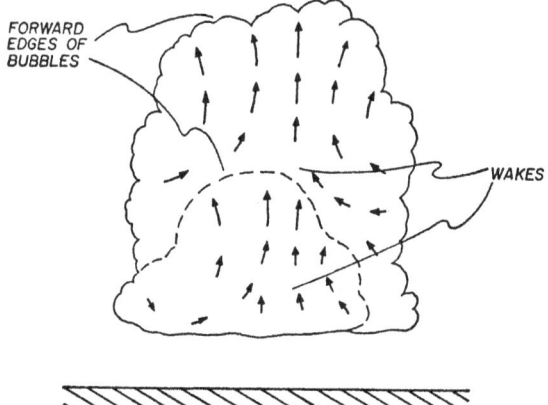

FIG. 12. *The bubble theory pictures a convective cloud not as a single column of rising air but, as shown here, as a succession of discrete bubbles.*

The new theory specifies that several maxima of updraft, temperature, and water content may exist. Within each bubble these quantities increase with altitude, but between bubbles the vertical motions are relatively slow, and the water content and temperature excess are low.

You might ask why the question of column or jet convection versus bubble convection has not been resolved. It would appear offhand that a few measurements taken at the right place and time would settle the question. Of course they would, but so far the right kind of measurements have not been obtained. One of the chief reasons has been the lack of suitable equipment to cope with such a variable phenomenon. What we need are devices to measure the vertical distributions of temperature, cloud water content, and, if possible, vertical velocity of the air.

Most of the measurements of the properties inside the clouds have been made with airplanes. Unfortunately, planes are constrained to fly essentially horizontal traverses. We need vertical soundings through the clouds. If many horizontal cloud penetrations could be made simultaneously, one could probably reconstruct the cloud pattern, but so far data of this type have not been collected. As a result, meteorologists have two theories of thunderstorm structure. Each one can explain certain observed features of the clouds, but there are insufficient grounds for rejecting either one.

Rain and Hail from Thunderstorms

The heaviest rates of rain are observed to fall from thunderstorms. Also, hail is peculiar to these storms. Ice particles, such as sleet, do occur with other cloud types, but these ice particles are relatively small. When you mention hail, you think in terms of pieces of ice which, not infrequently, are as large as walnuts and sometimes exceed the size of baseballs.

It is not surprising that hail and heavy rain fall from the same clouds. Both elements require large cloud masses with high water contents and strong updrafts to create them.

When you measure rainfall rate at the ground, you measure the amount of water that arrives in a unit time, either an hour or a day. The rate is a measure of the mass of the individual raindrops and the speed at which they approach the ground.

There are various ways in which small cloud droplets or ice crystals may begin to grow, but once they have attained diameters of about 100 microns (.004 inch) by far the most important growth mechanism is collision and coalescence. Large, sparse drops (e.g., 100 per cubic yard) falling faster than the many more smaller cloud droplets (e.g., 1500 per cubic inch) collide with and merge with them. The enlarged drops then fall and grow even faster. In this way raindrops of the order of .04 inch in diameter can be formed in a short time, if they remain in the cloud long enough. A drop of this size has a terminal speed in still air of about 700 feet per minute. In order for the drop to remain very long in a cloud, the cloud air must be moving upward at a faster rate. As long as the drops are kept inside the cloud, they will continue growing.

Once large drops have been formed, they can make their contribution to heavy rain intensities by falling to the ground rapidly. This occurs when they are carried toward the ground in downdrafts. Measurements by aircraft show that the air may descend at speeds of several thousand feet per minute. When this is combined with the fall speeds of the drops in still air, the drops can be moving at speeds exceeding a mile per minute relative to the ground. In such a case, the number of large drops striking the earth per unit of time is quite high and torrential rains result.

Hailstorms, because they may cause great damage and because they are not too common, have always been of great interest. The exact mechanisms by which they form are not yet clearly understood, but there are some areas of agreement among the experts. From observations at the ground, we know that large stones are composed of layers of relatively clear and opaque ice. Any theory of hail formation must explain these laminations as well as the extreme sizes (Plate III).

An item of pertinent information which has been known for a long time is that when a convective cloud builds to altitudes where the temperatures are much colder than freezing, the cloud droplets usually do *not* freeze until very low temperatures are reached. At 14° F to – 4° F, the cloud droplets usually are still liquid. Liquid droplets with temperatures as low as – 39° F are possible. In this state, the droplets are said to be *supercooled* or *subcooled*. They will remain in liquid form until they come into contact with ice crystals or special types of particles, known as ice nuclei. When this happens, they begin to freeze.

If an ice crystal or a large frozen cloud drop falls through a region of supercooled cloud droplets, it may grow into a larger ice particle as the intercepted droplets freeze. If the updraft speeds are small, the particle will soon fall out of the cloud, melt when it reaches warm temperatures, and reach the ground as rain. The largest raindrops from a thunderstorm usually are melted ice particles. If, on the other hand, the updraft speeds are high, the frozen particle will be supported in the cloud and may continue growing at an ever-increasing rate until it becomes a respectable hailstone.

What about the layers of clear and white ice? At one time it was thought that these laminations were produced by excursions above and below the freezing level. It was conceived that the only way to obtain clear ice was to permit the ice particle to start to melt and develop a layer of water, which would freeze again when a stronger updraft carried it to colder regions.

It is now known that clear ice can be developed in the subfreezing part of the cloud if the liquid water intercepted is sufficiently copious. Assume that the falling hailstone collides with a large number of supercooled drops. In order for them to freeze almost instantly and thus produce opaque ice, the heat of fusion must be extracted at an extremely rapid rate. An ice cube in a glass of water takes a long time to melt because the necessary heat is supplied slowly. The reverse problem, that of freezing, is in large measure analogous. Since the supercooled water is not rapidly frozen, it collects, runs over the stone, and freezes slowly. In this way a layer of clear ice is formed.

If the stone then falls into a layer of low cloud water content and few droplets are intercepted, the heat dissipation requirements can be met and the supercooled droplets freeze without spreading over the stone. Air bubbles are trapped in ice, and opaque ice results.

A single trip to the top of a tall thunderstorm and down again will permit the interception of enough supercooled water to account for ice particles of the order of half an inch. But for hailstones of diameters of 3 inches, a single continuous trip is not sufficient. The fall velocity of a 1-inch hailstone is about 4000 feet per minute. Such a stone would fall rapidly out of a cloud and would not have a chance to grow to the extreme sizes sometimes observed, unless strong updrafts were present to keep it in the cold part of the cloud. F. H. Ludlam has proposed a mechanism by which hailstones can be maintained in the subfreezing parts of the clouds for periods long enough to permit them to attain sizes of perhaps 3 inches and at the same time produce the laminations. He visualizes a process in which the stone falls out of an updraft which is caused to tilt because the winds increase with height. The stone falls through a region of low water content before it re-enters another portion of the cloud whose upward speed is enough not only to support it but to carry it upward a certain distance before it repeats the same process. There is high cloud water content in the main updraft and little in the region where the hailstone falls, and hence the onionskin appearance can be explained. Since the fall speed of a 3-inch hailstone is about 6000 feet per minute, it is evident that a cloud producing large hailstones must have very strong updrafts.

Lightning

By definition a thunderstorm is a storm from which thunder is heard. Where there is thunder there is lightning. Sometimes it cannot be seen, but you can be sure it occurred. There is little general agreement about the causes of lightning. To illustrate the degree of uncertainty in the whole subject, I might mention that there even has arisen a question whether Benjamin Franklin was the first to fly a kite and discover that lightning is an electrical phenomenon. The Russians have maintained that their own great scientist Mikhail V. Lomonosov preceded Franklin. It appears that Lomonosov, in fact, was conducting such experiments at about the same time. For the maintenance of international harmony perhaps it should be agreed that their discoveries were simultaneous.

At any rate, it is very well established that a lightning stroke represents a huge spark or arc between centers of differing electric charges. When the electric potential between two regions of a cloud or between a cloud and the ground exceeds the breakdown potential of about 25,000 volts per inch, a discharge occurs. Many studies of cloud-to-ground lightning have been made in various parts of the world. It is found that from the cloud there proceeds an avalanche of electric charges which moves downward perhaps 300 feet, slows down and then proceeds another 300 feet. By a series of steps the leading edge of the cloud of charges approaches the ground. When the forward tip of this so-called *step-leader* gets close to the ground, there is a sudden strong surge of charge up the same path. This surge is called the *return stroke* and is what you see. In a few thousandths of a second, tens of thousands of amperes of current flow up the channel. Once a path has been established, there may be many return strokes in rapid succession at intervals of a few thousandths of a second. The resulting ionization produces the blinding flash so characteristic of nearby lightning when seen at night. The heat created by such a strong current heats up the air and causes a sudden expansion, which produces a sound wave. When it passes overhead, the observer hears thunder.

Since the speed of light is 186,000 miles per second, the lightning flash is seen at almost the instant it happens. On the other hand, sound travels quite slowly, about 1000 feet per second. If the number of seconds elapsed between the sighting of the flash and the hearing of thunder is multiplied by 1000 feet, you can quickly obtain an estimate of the distance to the lightning flash. In general, lightning more than about five or six miles away usually is too far for the thunder to be heard.

An important gap in our understanding of thunderstorm electricity is a firm knowledge of the process or processes by which the huge magnitudes of electric charge necessary for lightning can be created. There are many theories, but none has been generally accepted. It is known that the upper parts of thunderstorms are positively charged, while the central parts are predominantly negative. At times, a small positive charge center is sometimes found in the rain in the lower part of the cloud (Fig. 13). The region of maximum electric field strength is between the main positive and negative centers, and it is in this region that most lightning strokes to airplanes occur.

The various theories of thunderstorm electrification can be separated into those requiring ice particles and precipitation and those which do not. Most meteorologists are of the opinion that the first category of theories is correct because lightning is usually not observed until fairly large clouds with ice in the upper layers have developed.

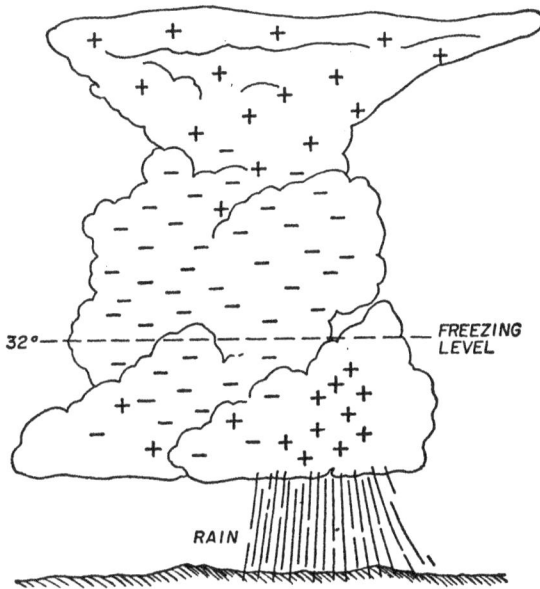

FIG. 13. *Distribution of electric charge in thunderstorms is shown here in relation to altitude and freezing level.*

Also, laboratory experiments, particularly at the New Mexico Institute of Mining and Technology under the leadership of E. J. Workman, have clearly demonstrated the role that ice particles can play in cloud electrification. It has been shown that when dilute solutions of water are frozen, a large electric potential develops between the water and the ice. The laboratory experiments show that the ice gains a negative charge while the water retains a positive charge. It is visualized that the proper circumstances for the creation of thunderstorm charge centers arise when a hailstone picks up more water than can be immediately frozen. After the freezing has started, some of the water is torn away by the air flowing around the stone. The small water droplets are carried upward to produce the positively charged upper part of the cloud, while the large ice particles fall toward lower altitudes. This process accounts for the main features of the charge distribution shown in Fig. 13. Laboratory experiments have also shown that friction between two ice particles causes a transfer of charge if one of the ice particles is warmed more than the other.

Other mechanisms involving precipitation have also been seriously considered as important in thunderstorm electricity.

One of the oldest ideas about the generation of thunderstorm electricity was presented by G. C. Simpson, who showed that the rupture of a water drop in a strong air current causes a separation of electric charge. In this process the large water particles are left with a positive charge while the air is left with a negative charge. This separation would lead to a polarity opposite to that associated with the main charge centers of thunderstorms. However, it is in the proper direction to explain the small positive center near the base of the cloud.

Several prominent scientists maintain that precipitation and, in particular, ice particles are not needed for the formation of large thunderstorm charge centers. Among the leaders in this group have been R. Gunn and B. Vonnegut. Their theories differ in principle, but neither requires the presence of ice particles. They rely on the capture of ions, minute charged particles in the air, by cloud droplets. Over the last forty years many theories involving this idea have appeared in the literature. Some of the early work was done by the great English scientist C. T. R. Wilson. The variations of the so-called ion-capture theories are many, and there is laboratory evidence that some of them are effective. One of the strong arguments made by their supporters is that there have been observations of lightning in small convective clouds when there was no ice. If these observations can be shown to be accurate, then it is evident that ice particles are not necessary, and theories involving ion capture become more tenable.

Quite obviously there is a need of better observations of cloud electrification as a function of the cloud history. At the present time a number of groups in various parts of the world are pursuing this elusive problem.

Gustiness at the Surface

A very noticeable feature signaling the approach of a thunderstorm is the gustiness of the surface wind and the associated fall of temperature. These events can be explained as consequences of the thunderstorm downdraft. Air from high regions of the cloud descends because the precipitation particles evaporate and cool it. When it reaches the ground, it spreads outward. The descending air also retains a certain amount of the horizontal motion it had when it was in the high layers of the cloud where the winds were strong. Near the ground its speed still exceeds that of the surrounding air. The gusty winds, which sometimes exceed 50 miles per hour, are capable of causing considerable property damage. These winds are sometimes referred to as *plow winds*. The outward rushing air usually has a maximum component in the direction toward which the thunderstorm is moving. When you feel the cool blast of a thunderstorm gust, run for cover before the rain, lightning, and, possibly, the hail arrive.

Lines of Thunderstorms–Squall Lines

In the spring and summer many parts of the United States are visited not only by individual thunderstorms, but by lines of storms that extend for many hundreds of miles. When you view them from a particular location, it is impossible to appreciate the extent and organization of the thunderstorms, but with a radar set which can simultaneously observe precipitation over areas exceeding 50,000 square miles, you frequently see lines of thunderstorms (Plate VIII). By combining the observations of a network of stations, M. G. H. Ligda has shown that the lines sometimes extend more than 500 miles.

In general, the lines of thunderstorms follow the pattern described in Chapter 2. They tend to be oriented along the direction of the winds at some intermediate level in the clouds.

It has been said that the thunderstorms constituting part of a squall line are the most vicious of all. It is true that the maximum frequency of hail and, as will be seen in the next chapter, the maximum frequency of tornadoes, are associated with squall lines over the Great Plains. However, it has not been established with certainty that these squall line thunderstorms contain stronger updrafts or turbulence or reach greater altitudes than do the isolated thunderstorms. Nevertheless, they do constitute a great menace to life and property because of their large areas.

These lines usually are oriented roughly northeast-southwest, and airplanes flying across the country are frequently faced with the job of penetrating them. Circumnavigation is an easy solution when widely spaced thunderstorms occur on the route, but this is not practical when a line several hundred miles long lies across the flight path. As recently as ten years ago such a trip represented a hazardous undertaking. If not dangerous to life, it certainly caused many an unhappy and uncomfortable flight crew. Commercial airplanes are very sturdy and can take more punishment than most of us realize. The wings may flap, but they almost never come off. Lightning may make a few small scars but that is about all. Hail will make dents in the metal but seldom affects anything but the airlines mechanics who have to repair the skin. All these buffets will leave minor impressions on the airplane, but on the passengers who have been through such a flight the scars are sometimes more permanent.

Fortunately for all concerned, higher-flying airplanes and airborne radar, as a young hotshot might put it, have taken the fun out of thunderstorm flying. Commercial airplanes now carry radar sets, which will locate the areas of maximum thunderstorm turbulence and hail. With this information on hand, the pilots can guide their airplanes through squall lines with a maximum of safety and passenger comfort. That is not to say that occasionally a thunderstorm will not be encountered with its unpleasant consequences, but compared to yesteryears' flying, squall lines are no longer a major obstacle.

Because of the extensive nature of squall lines, they represent a serious flood hazard. The thunderstorms in the line, like all thunderstorms, are constantly building, raining, and dissipating. Occasionally thunderstorms form upwind of a particular place and in succession drop their rain as they move over this place. If this is repeated a number of times, tremendous quantities of rain may fall over a period of several hours. Quantities exceeding 10 inches have fallen in a single day in parts of Illinois, according to reports of the Illinois State Water Survey. Of course, the consequences of such quantities of water in such a short period of time are clear—flooded farms, rampaging streams and rivers.

The gustiness at the surface on the forward parts of squall lines sometimes is quite extreme. This violence may occur when several thunderstorms close to one another have simultaneous downdrafts. The cool outrushing air at times may attain sufficient velocity to blow over buildings, damage parked airplanes, and destroy crops.

The most dreaded feature of the squall line is that it is the spawning ground for tornadoes, the real killers of atmospheric disturbances. Their properties will engage us in the next chapter.

CHAPTER 5

TORNADOES

The Midwest is a wonderful place in the spring-
time as the signs of winter disappear and the dor-
mant plants spring back to life. But when the western sky becomes covered with thunderstorms and
the rumble of thunder changes to distinct crashes,
it's time to look toward the storm cellar. The line
of thunderstorms could be bringing with it the violent vortex known as the tornado. That there is
good reason for fear is apparent in the following
account and a few statistics.

> *"One of the injured may have had the ex-
> perience of being in the center of a tornado
> funnel. The following is an account of his ex-
> perience. He reports that while driving west
> on the highway east of Scottsbluff, he noticed
> a large dust cloud but he did not recognize
> anything ominous, having observed many
> more awesome dust clouds in the past; so he
> proceeded to drive into the dust cloud at a
> point 3 miles east of Scottsbluff. On entering
> the dust cloud he realized this was no ordi-*

nary disturbance and stopped the car at the side of the road. There was a roar and a crash of glass as the windshield and windows were broken by flying debris. He pulled his wife's head over in his lap and bent over to shield their faces. There was a moment of comparative calm and he raised his head to peer through the broken windshield. Large boards, tree limbs, and a boulder the size of a man's head were floating around the car. When asked the direction of movement he stated without hesitation that the debris was circulating to the left or counterclockwise. Time here was indefinite until there was a crash and that is all he remembers until he regained consciousness in a hospital. Actually, both occupants were thrown from the car into the highway. Both were badly cut and torn. The wife was apparently killed instantly. The car was rolled into an unshapen mass of metal and deposited in a nearby field."*

In the period from 1916 to 1953 an average of 230 people a year were killed by tornadoes. The number for individual years has ranged from as low as 36 in 1931 to as high as 842 in 1925. Property damage has averaged almost $14,000,-000 per year but in 1927 was as high as $25,000,-000 in the state of Missouri alone. On the average, from 1916 to 1950, about 150 tornadoes have been reported each year with the numbers ranging from 65 to 220. It should be realized that these

* From article by E. V. Van Tassel, *Monthly Weather Review,* U. S. Weather Bureau, Washington, D.C., Vol. 83, November 1955, p. 255.

numbers are not very accurate. Very few of the reported tornadoes are seen by regular weather observers. Tornadoes are small and short-lived. The Weather Bureau file on tornadoes is based largely on newspaper accounts and information supplied by people who happened to be in the vicinity of the tornado. It is evident that as the population density over the Great Plains has increased, the number of tornadoes reported each year has shown an upward trend. The average of 150 per year certainly should be considered to be much lower than the true value.

Tornadoes have been observed in all parts of the world and every one of the continental states, including Alaska, but are largely concentrated over the Great Plains and Middle West. Leading the list are Kansas, Iowa, Texas, Oklahoma, and Arkansas. Why they form in this region will be discussed later. At this point it is appropriate to describe the tornado and show why it is so destructive.

Description of a Tornado

Tornadoes are quite small; the vast majority are less than a mile in diameter and many are less than 100 yards. They appear as pendent funnels which dip downward from the base of existing clouds and approach the ground in an irregular fashion. Many times funnels are seen that never reach the ground at all. Instead, they oscillate downward and upward several times and finally disappear in the clouds.

The appearance of a fully developed tornado, or *twister* as it is sometimes called, can have a variety of shapes (Plate IV). Sometimes they look

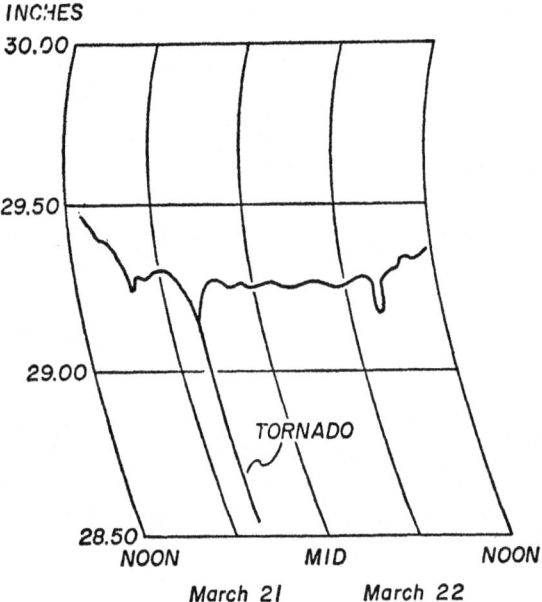

FIG. 14. *The characteristic low pressure at the center of a tornado shows up in this barograph of a storm that passed Dyersburg, Tennessee, at 8:50* P.M. *on March 21, 1952. A record like this is a rarity because storms seldom pass close to barographs and may destroy all the instruments in their paths. From J. A. Carr,* Monthly Weather Review, *Vol. 80, 1952.*

very much like an ordinary funnel, wide at the top and tapering to a small diameter at the earth's surface. At times the tornado is a large circular cylinder whose diameter changes little between the cloud base and the ground. Occasionally the funnel has the appearance of a long rope, very narrow and twisted in peculiar shapes, sometimes even having a large horizontal section.

The outer boundaries of many tornado funnels are very distinct. Other times the tornado is a fuzzy mass of clouds and dust.

One common feature of all tornadoes is the low pressure at the center of the storm (Fig. 14). Because the strong winds usually destroy instruments in their path and because it is not very probable that small funnels will move over the very widely spaced barographs, measurements of the pressure at the centers of tornadoes are very few. Even those that do exist are subject to question because most barographs are not designed to respond to the fast pressure variations produced by a moving tornado.

There have been measurements showing rapid falls of about one half an inch of mercury followed by an equally rapid rise to almost the value prevailing earlier. The Weather Bureau has cited a tornado in St. Louis in 1896 in which a pressure difference of 2.42 inches of mercury was observed over a distance of seven blocks. It is to be noted that a drop of 2.42 inches represents a sudden reduction of normal atmospheric pressure by about 8 per cent. Because of the limitations of the response time of the barographs employed, it is reasonable to expect substantially larger pressure falls than those reported.

It is important to recognize that the pressure changes occur very quickly. As an example, with a tornado having a diameter of 500 yards and moving at a speed of 20 miles per hour, the entire system would move over a point in less than one minute. This means that the pressure would drop to its minimum value in about 30 seconds.

The tornado funnel one sees is actually a cloud of water droplets mixed with dust and other debris. Close to the ground the dust and debris are plentiful because the low pressure causes air to flow inward and upward. Close to the ground one frequently sees a scarf of dust and other material stirred up and thrown outward from the region of high wind speeds. Away from the ground water droplets cause the outline of the funnel cloud. The droplets form because the low pressure in the funnel causes an expansion and cooling of the air, which lead to an increase of the relative humidity and condensation. This is the sequence of events occurring when a parcel of air ascends from low to high altitudes (and lower pressures) and forms a cloud. In the tornado the large pressure differences at the same altitude cause clouds to form in the regions of low pressure.

Probably the most striking feature of a tornado to anyone who has been close enough to see one in action is the velocity of the wind. The maximum speeds have never been recorded because the anemometers in a position to make the measurements have never survived. Speeds as high as 120 miles per hour have been measured, but to cause the types of damage observed, much higher velocities would be needed. Many authorities have suggested that the winds could exceed 400 miles per hour. One has estimated that winds in localized regions of the funnel may reach peak speeds close to the speed of sound.

The rotation of the winds in tornadoes is almost always counterclockwise, but there are some re-

PLATE I. Small cumulus clouds form in lines along the wind direction near Tampa, Florida. The photograph was taken at 30,000 feet. [Courtesy V. Plank, Air Force Cambridge Research Center.]

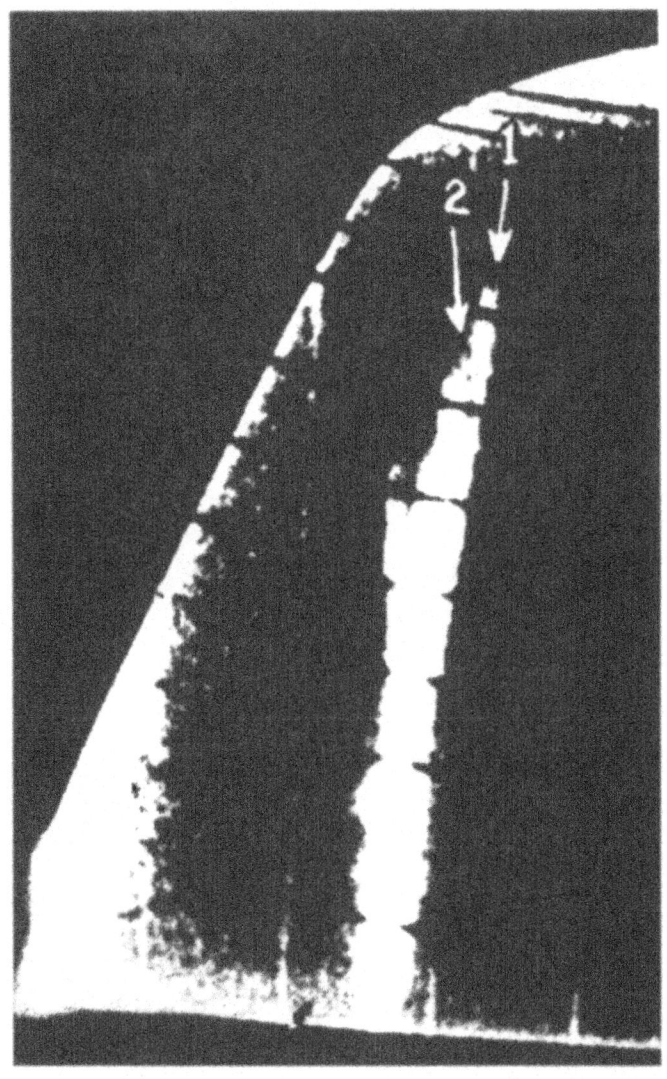

PLATE II. Radar echoes from a large thunderstorm formed this picture on the radar screen twenty-seven miles away. A storm's large water and ice particles reflect the radar waves. The storm tower marked 1 extended to about 45,000 feet, tower 2 to 38,000 feet.

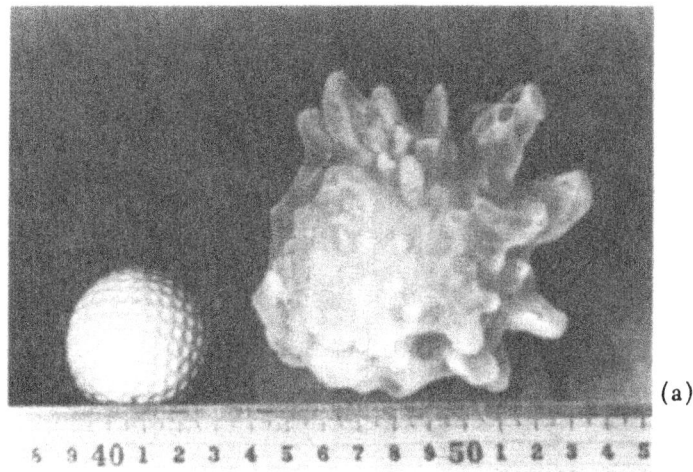

(a)

PLATE III. The magnitude of large hailstones is evident in this comparison (a) with a golf ball and meter stick. [Courtesy U.S. Weather Bureau and W. Wise.] The crystal structure of a hailstone is photographed in (b) with polarized light. Under ordinary light the bands of small crystals would look like opaque ice, the large crystals like clear ice. This slice of hailstone is about two inches in diameter, the millimeter (25 mm equal one inch) being the scale unit. [Courtesy V. J. Schaefer.]

(b)

(a)

PLATE IV. Tornado funnels assume various shapes, as these photographs show. [Courtesy (a) Rudd Studio, Dallas, Texas, and U.S. Weather Bureau; (b) C. B. Raymond, Dallas, Texas, and the U.S. Weather Bureau; (c) C. Kavanaugh, Taylor Publishing Company, Dallas, and U.S. Weather Bureau; (d) U.S. Weather Bureau.]

(b)

(c)

(d)

(a)

PLATE V. Like tornadoes, waterspouts take various shapes. [Courtesy (a) A. O. Bliss and U.S. Weather Bureau; (b) John Lambert III and U.S. Weather Bureau; (c) U.S. Weather Bureau and Air Force.]

(b)

(c)

PLATE VI. On June 9, 1923, a tornado roared across Westboro, Massachusetts, leaving the damage shown in this aerial view. [Courtesy Worcester *Telegram-Gazette* and U.S. Weather Bureau.]

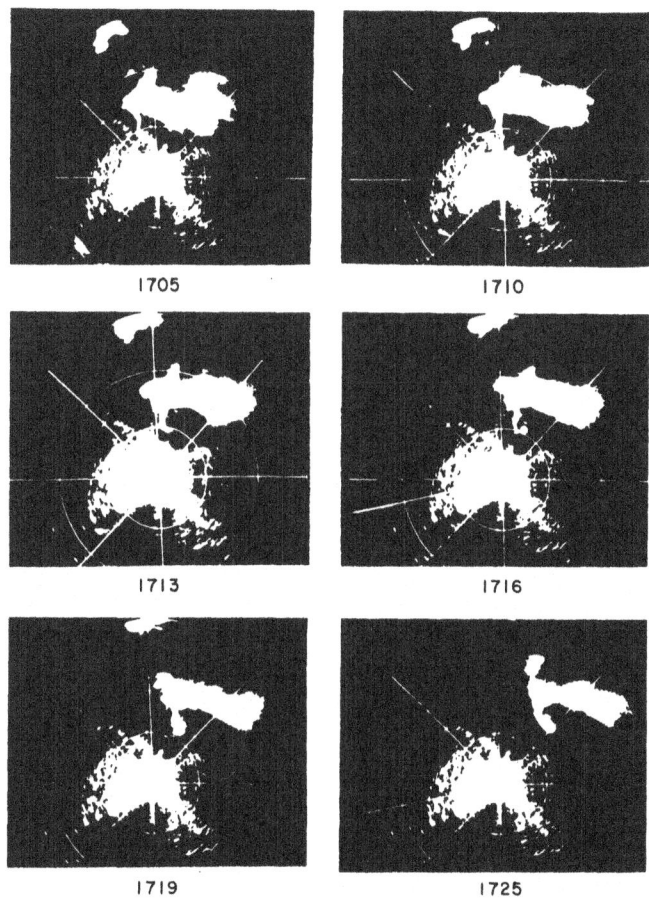

1705

1710

1713

1716

1719

1725

PLATE VII. Formation of a tornado "hook" can be followed in this sequence of radar observations, the first ever made of a tornado. Reflections from the ground caused the large bright center and the small dots. The tornado occurred in the most southerly portion of the protruding finger. The time is shown in hours and minutes according to the 24-hour clock. [From G. E. Stout and F. A. Huff, *Bulletin American Meteorological Society,* Vol. 34, 1953.]

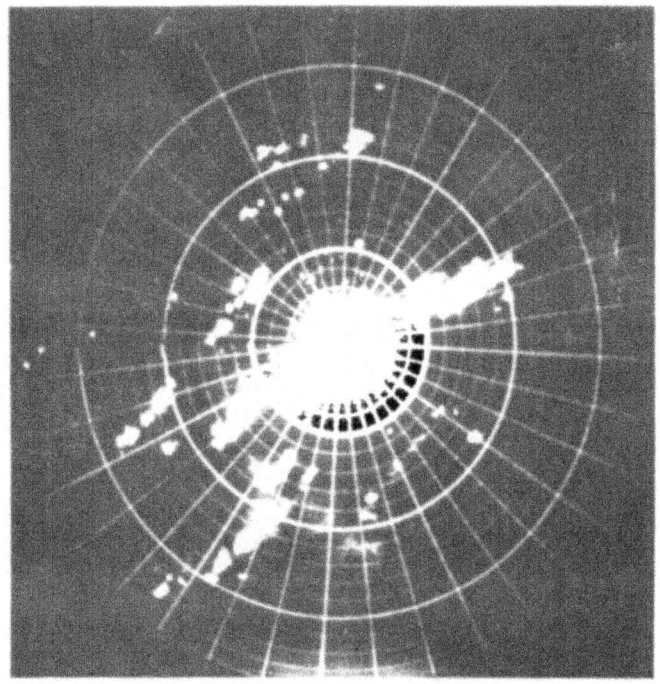

PLATE VIII. Lines of thunderstorms show up on radar. The heavy concentric rings are 50 miles apart. The bright images show thunderstorms in a generally northeast-southwest alignment.

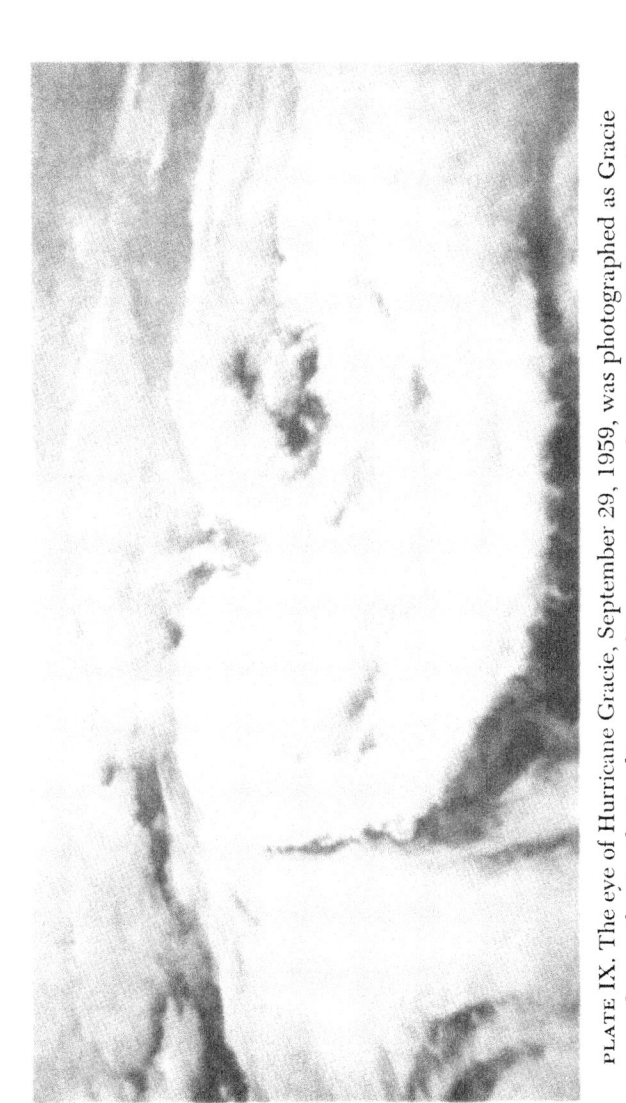

PLATE IX. The eye of Hurricane Gracie, September 29, 1959, was photographed as Gracie passed over the South Carolina coast. [Courtesy Commander R. S. Hill, U.S. Navy, official photograph.]

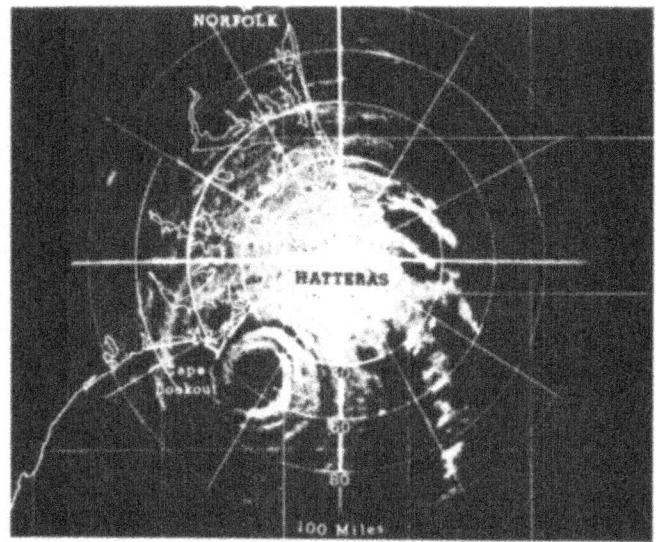

PLATE X. Spiral bands of precipitation appear in this radar observation of a hurricane. The radar station was at Cape Hatteras. A map of coastal North Carolina and Virginia is superimposed. [Courtesy U.S. Weather Bureau.]

PLATE XI. Circular or spiral cloud bands were seen in the eye of a hurricane on at least this occasion. [Courtesy Commander R. S. Hill, U.S. Navy, official photograph.]

ports that they sometimes spin in the opposite direction. In this connection the tornado vortex differs from the dust devil vortex because the latter frequently spins clockwise.

A characteristic feature of tornadoes is the noise associated with them. Almost invariably people who have been unfortunate enough to have been in the path of a tornado have commented that the storm approached with a great roar. Such comments as "a thousand railway trains," or "the buzzing of a million bees," or more recently the "roar of flights of jet airplanes" are common. The cause of the noise is still a mystery, but a number of speculations have been offered.

One explanation has been that in parts of the vortex the winds may attain sonic velocities and produce shock waves of small amplitude which will be heard as a roar. Another explanation, particularly of the buzzing or hissing sound, is that it is a result of electrical discharges or smaller whirls inside the tornado. In some tornadoes the electrical displays have been spectacular. An unusual view of the inside of a tornado was obtained by Mr. Will Keller, a farmer who observed a tornado as it passed over his storm cellar.* He reported that the tornado center was hollow, about 50 to 100 feet in diameter, and that the walls of the vortex were lighted by constant lightning flashes which "zigzagged from side to side." He also stated that smaller vortices were forming along the walls and moving around the inside of the main vortex and then breaking away. They looked like

* See article by A. A. Justice, *Monthly Weather Review*, U. S. Weather Bureau, Washington, D.C., Vol. 58, May 1930, p. 205.

tails as they writhed their way around the end of the funnel. It was these, he said, that made the hissing noise.

Mr. Keller reported that the inside of the tornado was hollow and appeared to extend upward at least one half mile. This feature has also been mentioned by at least two other witnesses who looked a tornado in the eye and lived to tell about it. Other observers have also suspected that the inner parts of the funnel are free of cloud, because when viewed from the side, tornadoes sometimes look less dense along the central axes. If the interiors of funnels are indeed hollow, this is quite significant. It may mean that the air in this region is comparatively dry and moving downward. At one time it was commonly assumed that the motion was upward throughout the funnel. Some scientists now speculate that the rising air is restricted to a shell representing the outer boundary of the tornado.

Why Tornadoes Are So Destructive

A tornado destroys property and causes loss of life because the low pressures lead to the explosion of closed buildings and vehicles and the strong winds blow away whatever lies in their path.

The reasons for the explosions are well known. As we said earlier, the pressure in a tornado may cause a drop of atmospheric pressure by 8 per cent or more in a matter of seconds. Suppose the pressure inside a house is normal atmospheric, about 15 pounds per square inch. If a tornado moves over the house the pressure outside may suddenly drop by 8 per cent to 13.8 pounds per square inch. Since the pressure inside the house will drop fairly slowly, especially if all the doors and windows are closed, the force on each square inch of wall and ceiling may amount to 1.2 pounds or about 170 pounds per square foot. If the house had a ceiling space 20 by 40 feet in area, the force exerted on the roof would be about 68 tons. This suddenly applied force can blow the roof off the house as if an explosion had occurred. This is especially true for dwellings, because in most houses the roof is held on mainly by its own weight. And it is evident, too, that few walls will survive 170 pounds of force per square foot. Often a tornado will cause several walls and the ceiling to blow out simultaneously.

It should be noted that these calculations were based on an 8 per cent drop in pressure, a fairly conservative maximum amount. In some cases the pressure drop and consequently the explosive forces could be considerably larger.

The explosive effects of tornadoes lead to many reports such as the following taken from the 1956 *Climatological Data,* prepared by the U. S. Weather Bureau:

". . . house simply fell apart when tornado hit."

"Roof of high school lifted and building declared unsafe."

The pressure differences between the inside and outside of buildings or vehicles are greater when the windows and doors are closed. Unfortunately, squally winds with heavy rain, and very frequently hail usually precede tornadoes. In order to prevent rain from blowing into their homes and damaging the interior, rugs, furniture, etc., too many people close the windows and doors. They do not realize that they are risking far worse damage.

Occasionally the pressure effects are not severe enough to cause an explosion. Perhaps the windows were left open to permit the air inside to rush out and allow the pressure inside to change rapidly. One source of danger has been avoided, but the occupants still are far from being out of danger. The strong winds associated with the tornado are capable of picking up and moving the entire house. Cars, trucks, trailers, and other heavy objects are frequently carried away.

Reports like the following are not at all unusual: *

"8-room house lifted off its foundation, leaving 19 members of family inside, unhurt."

"Demolished 1 hangar, lifted another 4 feet off ground and set it back in place. Did not damage plane in demolished hangar."

"Car picked off highway, whirled around several times and deposited in field 250 feet north of road. Driver suffered only shock and some bruises. Car trunk blown open with lid bent over car top."

"Long-span steel bridge blown off its supporting abutments. A deep freeze carried ½ mile. Many structures burst outward as tornado passed."

In the four cases cited here, there was no fatal-

* From *Climatological Data*, 1956 and 1957, prepared by U. S. Weather Bureau.

ity. Many times the occupants of the buildings or vehicles are not so lucky.

The most damage and loss of life occur when the explosions and the winds combine their lethal powers.

Eyewitness accounts as well as motion pictures have shown that, frequently, the remnants of what was once a house are picked up by the strong winds and blown through the air like huge pieces of shrapnel. Whatever gets in the way is cut down. A large fraction of the fatalities associated with tornadoes are caused by debris, huge timbers, pieces of metal, glass, etc. The evasive action one should take in this situation is to find the lowest spot and lie as low as possible with head covered. The dangers are the same as those faced by soldiers under bombardment. The only relatively safe place is in a hole.

The combination of blow-out and strong winds can level a town like a huge bomb. Aerial photographs show that large sections of towns have been completely wiped out with only the building foundations left intact (Plate VI).

Some Unusual Tornado Stories

Weather Bureau records and newspaper files contain tornado stories that are truly astounding.

Reports of thin pieces of straw blown into tree trunks or fence posts are numerous. S. D. Flora in his book *Tornadoes of the United States* cites the case of a two-by-four plank blown through solid iron sheet five-eighths of an inch thick. Incidents like these attest to the tremendous wind velocities involved.

There have been many stories of turkeys and chickens stripped clean of their feathers. Usually the fowls lose their lives as well as their plumage. But at times they have been found barely alive. In a study of this problem G. W. Reynolds concluded that the feathers were removed by "blowing off" by strong winds rather than by being "blown out" because of pressure effects.

Tornadoes sometimes play havoc with fences. They may pull out the posts and roll up the wire as if it were knitting yarn. One report said that "fences were rolled into 50-foot diameter balls."

Many times tornadoes pick up things and carry them many miles before depositing them, sometimes with a crash, other times quite gently.

The following reports are examples taken from Weather Bureau publications. They illustrate the unusual and tragic consequences of a tornado.

"After tornado, packages of knitting products from wrecked knitting mill were recovered undisturbed 35 miles north of where tornado occurred."

"Boy picked up by wind and carried 40 feet, not seriously hurt but terrified."

"Some unusual features were: a government bond found 60 miles southeast of Eldorado address, eight $100 bills found intact in envelope far from owner's home in Eldorado, boy found with dozen splintered sticks protruding from his chest, woman sucked through window and blown 60 feet from house, and beside her was found a broken record entitled 'Stormy Weather,' auto-

*mobile carried more than a block and
jammed through roof between a bed and a
dresser . . ."*

A very interesting account of an experience with
a tornado is reported in the E. V. Van Tassel article
cited earlier. Three broadcasters were observing a
tornado from a mobile unit and giving a blow-
by-blow account.

*"These broadcasters were ahead of the funnel
and to maintain distance they drove through
a west gate of a cemetery intending to depart
through a south gate. As fate would have it,
the south gate was heavily chained and
locked. Escape was cut off by the tornado
moving over the west gate. The three broadcasters abandoned the mobile unit for the
basement sanctuary of a stone building, but
left the mobile unit in operation so that the
noise of the tornado was broadcast to the
public from a distance of not more than 100
feet from the very center of the tornado. This
noise was a very audible roar that might be
compared to many trains passing in unison.
Though the automobile housing the mobile
broadcasting unit was damaged to the extent
of $1200 it remained upright and the broad-
casting unit continued to function through the
entire period. The broadcasters in the basement or furnace room were having the experience of a lifetime. As they huddled
around the furnace they observed tools, such
as shovels, hoes, rakes, etc., scoot up the en-
trance ramp and disappear. Then came total
darkness and a deepening roar. The furnace*

*twisted and heaved and the broadcasters
found it difficult to breathe. Whether this was
due to pressure or lack of such could not be
determined. Time in this case seemed to be
in minutes and over this time the temperature
dropped from a mild summer value until the
broadcasters were chilled until they were actually cold. The roar moved on east, light returned, and the broadcasters emerged unhurt
to observe the tornado funnel in action continue a slow movement to the east as they
resumed voice broadcasting describing the
location and action of the tornado funnel.
One observation was of large pieces of debris
rotating counterclockwise in the funnel and
slowly going upward and then suddenly dropping rapidly toward the ground. These large
pieces would stop just short of the ground
and then again rise slowly while rotating."*

The Formation of Tornadoes

Although meteorologists still do not know how
and why tornado funnels form where they do, they
can specify the conditions usually associated with
their development. Many speculations have been
offered to explain the processes involved.

Tornadoes in the United States are most common in the spring and early summer, but they may
occur at any time. It is known that in the early
part of the year, March and April, they are more
common near the Gulf Coast. As the year pro-
gresses, the center of the region of maximum tornado likelihood moves northward. By June the
greatest tornado risks are in Kansas, Nebraska, and

Iowa. The statistics show that Texas and Oklahoma are relatively free at these times. The reason for the shift northward is tied to the movement northward of the air masses favorable for tornado development.

It has been known for a long time that prior to the occurrence of tornadoes the atmosphere consists of a deep dry layer of air on top of a moist layer. The humid air originates over the tropical oceans. In the United States the inflow of moist air below 10,000 feet is seen on the weather maps as a fairly strong stream coming out of the Gulf of Mexico.

The upper dry layer is composed of air that has passed over the Rocky Mountains and been subjected to some sinking motion. At the boundary between dry and moist regions the temperature increases with height. This *inversion* of the normal lapse rate of temperature shows the presence of a stable layer tending to suppress small-scale thermals. As far as individual parcels of air are concerned, this region of the atmosphere is stable. However, if some mechanism causes extensive lifting of a very large mass of air, it will become quite unstable. This occurs because the moist air will reach saturation before the dry air. As a result the lower air will start cooling at the smaller moist adiabatic rate sooner than will the upper air, as you will recall from Chapter 2. If this process continues long enough, the lapse rate of the large mass of air will become unstable. In this case the air is said to be *convectively* unstable.

Until the dry-over-moist air mass is lifted en masse, thunderstorms and tornadoes are not likely. In addition to the lack of thermal stability, the dry air aloft will act to reduce the buoyancy of the rising parcels. Mixing of environment air into the cloud through its top and sides will cause evaporation and cooling.

There are very few observations that actually show the modification process by which an air mass with a stable layer and dry air aloft is converted to one with an unstable lapse rate and a deep moist layer. It appears that this is a fairly local effect occurring in the general vicinity of the storms. From an examination of 2465 soundings of the atmosphere R. Beebe found only eleven cases where suitable observations were taken within 50 miles and less than one hour prior to the occurrence of a tornado. This scarcity of information is not odd because the radiosonde stations which release balloon-borne instruments to measure temperature, humidity, and pressure are several hundred miles apart and usually take only two or four observations per day. A research program now in progress under the direction of the U. S. Weather Bureau is employing specially instrumented airplanes to accumulate measurements which should shed light on this important problem.

Weather forecasters of the U. S. Air Force and the U. S. Weather Bureau have found that an important feature of the weather pattern at the time tornadoes form is a jet of strong winds at intermediate levels of the atmosphere, somewhere in the vicinity of 15,000 feet. The lines of thunderstorms which usually form prior to the tornadoes tend to be roughly parallel to these winds. The wind distribution and the thunderstorms appear to be the keys to the explanation of the bodily lifting of the air mass. Under certain conditions the airflow will cause the air to converge in the lower layer of the atmosphere and to diverge in the superposed upper layers. The result is ascent of a large body of air. Although the vertical velocities can be relatively low, say 2 inches per second, if these velocities continued for 12 hours, the air could be lifted more than 7000 feet.

When the air is convectively unstable to a marked degree and there is a strong jet, conditions are ripe for thunderstorms, hail, and tornadoes, in that order. Many thunderstorms are observed, some produce hail, and a very few produce tornadoes. There are various hypotheses why only a few thunderstorms produce tornadoes and how they do it. None of the present hypotheses is generally accepted by meteorologists. By the same token a few of them cannot be rejected.

The hypotheses that have been seriously considered have tried to show how the vortex and localized upward motion are started. Various scientists have argued that the tornado originates as a region of circulating air in the thunderstorms which works its way toward the ground. Most observations of tornadoes do show that the funnels originate in the clouds and descend toward the ground.

Pressure observations at the ground made by E. M. Brooks of St. Louis University have shown that tornadoes frequently occur within a small region of low pressure. This area, perhaps 5 miles in diameter, known as a *tornado cyclone*, has a slow counterclockwise rotation. It has been suggested that the much smaller tornado results from a concentration of rotating motion which is accompanied by upward motion. The mechanism for the concentration has not been specified.

Recently B. Vonnegut has proposed that tornadoes are initiated and maintained by energy supplied through lightning discharges. He has cited the fact that many tornadoes produce unusual electrical effects. On the other hand, it is known that intense lightning is a common occurrence of many thunderstorms that never produce tornadoes. Although this hypothesis has not had much support from meteorologists, it has some fascinating aspects.

Once the strong upward motion has been initiated and the tornado spin begins, it can be maintained by the addition of energy. The major source of energy is the heat released during condensation of cloud droplets. As air in the low levels rises, it is replaced by rapidly converging air. As pointed out in Chapter 3, the principle of the conservation of angular momentum demands that as the converging air gets closer to the center of rotation it must rotate faster and faster. In this way strong wind velocities are generated.

Some tornado systems involve a single funnel, others involve many of them. Some tornadoes last only seconds; others may go on for tens of minutes. There have been reports that some tornadoes have traveled as far as several hundred miles and lasted for many hours. It is questionable whether the same funnel could travel such a long path or whether there are a number of funnels forming in rapid succession but lasting only for a short time. This question is somewhat academic. It certainly is so to someone in the path of a tornado-producing thunderstorm.

Tornadoes may form at any time, day or night, but are most frequent in the afternoon. More than half (58 per cent) of the storms have been reported between the hours of 2:00 and 8:00 P.M., with 23 per cent between 4:00 and 6:00 P.M. It is not a coincidence that the maximum tornado frequency occurs just following the period of maximum temperatures at the ground. As we have mentioned in earlier chapters, the high temperatures contribute to instability and the formation of thunderstorms, which in turn can lead to the creation of tornadoes.

Waterspouts

When tornadoes occur over water, they are called "waterspouts" (Plate V). There is some evidence to suggest that, in general, they are not as violent as tornadoes over land, but still they have caused boats to capsize and damaged property when they moved over land. An interesting feature of the waterspout is that the existence of low pressure at the center can be seen by variations of the water level. A "mound" of water perhaps 2 feet high sometimes appears because the pressure in the water is higher than that of the air in the funnel and forces the water surface upward. Speaking loosely, one might say that the low pressure sucks the water upward.

The waterspout, contrary to the opinions held by many people, is not composed of water drawn up to cloud level by the low pressure and updrafts. Instead, as in a tornado, the visible part is composed of small water droplets formed by condensation. However, on some occasions the updrafts in the spouts have carried upward considerable quantities of salt-water spray created by strong surface winds. This has been verified by the fall of unusually salty rain water following the passage of a waterspout.

Although most spouts occur with convective clouds in a manner similar to the formation of a tornado, some have been seen with little or no cloudiness. It appears that they are similar, in some respects, to dust devils, which also occur with clear skies. As mentioned earlier, the dust devil indicates the presence of strong instability caused by intense heating of the earth's surface by the sun's rays. Waterspouts in clear skies must also indicate the presence of unstable air. The passage of such air over warm water, coupled with a generally circular air motion, probably sets up the vortex. Relatively little is known about the exact mechanism.

Tornado Detection and Protective Measures

At the present time the U. S. Weather Bureau issues alerts and warnings of the possibility of tornadoes. These alerts cover fairly large areas, perhaps 100 miles on a side. Of course, even when a tornado does occur, only a very small fraction of the people in such an area will be affected. The reasons for not restricting the area are twofold: (1) the fact that the weather observations needed for making forecasts are taken at stations spaced 100 to 200 miles apart and (2) the limitations in our knowledge of the processes leading to tornado formation. The inability to forecast the time and location of tornadoes precisely makes it increasingly important to improve the procedures for detecting them.

In the last few years the Weather Bureau has assisted many communities to establish networks of voluntary observers. These people assume a responsibility to make observations and immediately report to an appropriate official any severe thunderstorm or tornado. This information is then used for alerting such agencies as police, radio and TV stations, Red Cross, etc.

In recent years radar has been playing an important role in the tornado detection and tracking networks. In March 1953 the Illinois State Water Survey observed a tornado on radar for the first time (Plate VII). The peculiar hook-shaped echo shown in the diagram is a good indicator that a tornado is present in the cloud. Unfortunately, many tornadoes occur without this distinctive protuberance. Nevertheless, once a tornado has been spotted by visual observers and reported, it can then be tracked by radar. There is time to warn communities in the path.

Radar sets operated by various government and private organizations are currently in use for tornado detection. The records show that many lives already have been saved. S. G. Bigler, at Texas A. & M. College, was credited with saving many lives when he warned school authorities that the radar pattern showed a severe thunderstorm or tornado would pass in the vicinity of a school just about the time the students were to be dismissed. The children were kept in school and precautions taken. The tornado passed harmlessly down the street in front of the school.

When the Weather Bureau issues an alert, people in the area should get set. They should keep the radio on, tuned to the local station or one of the civil defense stations, 640 kc or 1240 kc. If a tornado is sighted and reported, its location will be broadcast.

When thunderstorms are observed, keep your eyes open for any sign of a tornado funnel. If you see a tornado and have time, notify the telephone operator, police or radio station. Open windows on the leeward side of the house and run for cover.

At night or during heavy rain the funnel may not be seen, but the loud roar may be heard. When this occurs, there is no time to use the telephone. The funnel is about to strike. Take evasive action immediately!

The safest place when a tornado is approaching is a storm cellar. Many people in those parts of the country where tornadoes are frequent have built underground rooms. These so-called *cyclone cellars* range from caves supported by timbers to sturdily constructed concrete rooms covered with several feet of soil, with suitable air vents and stocked with water and supplies. The construction of such an underground refuge is relatively easy and well worth the time and effort. It is the only sure place for surviving a tornado.

If there is no storm cellar, the next best place is the southwest corner of the basement. Be sure to open windows and doors to allow rapid pressure accommodations. Tornadoes usually move from the southwest and if a house is ripped down, the pieces are most likely to fall into the northeast part of the basement. Crouch close to the floor, and if a mattress or a table or other similar covering is available, get under it. These measures are relatively good with a frame house which will tend to go with the wind. In a brick house, the corner of the basement nearest the tornado offers some protection but there is real danger of falling bricks.

If you have no basement, crouch low against an interior wall on the ground floor. If the entire house goes, there is little chance of escaping unhurt, but interior walls often remain intact when parts of the outside walls are carried away. Also, this procedure reduces somewhat the hazards from timbers, glass, and other debris flung about in the tornado. Again, if there is anything to cover you, use it.

If a tornado strikes unexpectedly, while you are sleeping for example, crawl under something and lie on the floor. A bed, mattress, table, or even just a blanket could mean the difference between life and death.

When a tornado approaches a school, the chances for a catastrophe are high. It is recommended that the pupils be made to sit on the floor along interior walls. Before evacuating the school, be sure the tornado is still far enough away so that there is time to move to a safer place. A ravine or a deep ditch would be safer than most schools if everyone lay down close to the ground.

When you are in an automobile, you can outrun a tornado providing you don't run out of road. Tornadoes usually move at speeds of the order of 20 to 30 miles per hour and very seldom exceed 50 miles per hour. If you cannot drive away from it, do not stay in the car. Get out, find a hole or ditch and lie low.

It is quite evident that the prime concern when a tornado approaches is to try to stay alive. Little can be done to save property. It is worse than worthless to board up the windows and doors as is done when hurricanes strike. Small items can be carried to a storm cellar if time permits, but it usually is not advisable to worry about earthly possessions when a tornado is getting closer. The attempt to save some property may cost your life.

A last word regarding safety measures seems appropriate. It is wise to plan ahead. At home and in schools particularly the course of action to be followed in the event a tornado threatens should be known *in advance*. Drills should be a regular practice. When the tornado strikes, everyone must be prepared.

CHAPTER 6

HURRICANES

At 4:30 in the morning on June 25, 1957, the
New Orleans Bureau office issued a bulletin that
a storm center had formed in the southwestern
Gulf of Mexico. The highest winds were only 35
to 40 miles per hour, and few people were particularly disturbed. But this was an unusual storm.
With amazing rapidity it intensified and moved
northward. On the morning of the twenty-seventh
it slammed over the coast near the Texas-Louisi-
ana border and devastated the state of Louisiana.
As the storm approached, the winds increased to
over 100 miles per hour, and the tides rose at a
rate of about one and a half feet per hour. Between
four and five o'clock in the afternoon the water
was more than 10 feet above sea level at Cameron,
Louisiana. Waves 4 to 5 feet high, with some peaks
over 8 feet, were superimposed on the high water.
The water washed over the land carrying away
everything in its path—houses, livestock, human
beings. In less than two days more than 500 people lost their lives; 40,000 to 50,000 cattle were
killed, mostly by drowning. Property damage was

estimated at between $150,000,000 and $200,
000,000. The storm which caused this tragedy was
a hurricane—the now famous Hurricane Audrey.

What Is a Hurricane?

A hurricane is an intense storm that forms over
a tropical ocean with the winds blowing counterclockwise* around a central calm area called the
eye. The recognized definition of a hurricane states
that it is a windstorm in which the winds reach at
least 74 miles per hour. Maximum speeds of 100
to 150 miles per hour are quite common in hurricanes striking coast lines. Speeds as high as 200
miles per hour have occurred according to estimates made from structural damage.

The strongest winds produced by nature are
found in tornadoes, but they are restricted to fairly
small areas, usually less than one mile. In hurricanes maximum winds at the earth's surface usually do not exceed 150 miles per hour, but they
may occur over areas many tens of miles in diameter. Tornadoes last for periods measured in
minutes, hurricanes have durations of days. Hence
the property damage done by a single hurricane
can be tragically high even when compared with
effects of a tornado.

Hurricanes are found in various parts of the
world and have several names. The storms that
form in the western North Pacific Ocean and strike
Japan with regularity are called *typhoons*. In the
northern part of the Indian Ocean they are known
as *cyclones*. In Australia they are sometimes re-

* The direction is clockwise in the Southern Hemisphere.

ferred to as *willy-willies*. In most other regions of the world the term hurricane is common. Regardless of the differences in name, the storms have the same causes and similar properties.

The number of hurricanes observed in the western Atlantic Ocean, Gulf of Mexico, or Caribbean Sea has ranged from as low as two in 1929 to as high as 21 in 1933. Most of them occurred in August, September, and October. In recent years, it has become common practice to give girls' names to the hurricanes in order to make identification a simple matter. The first hurricane of each summer is given a name beginning with an *A,* the second with a *B,* etc. Each year the names are changed. The hurricane is permanently identified both for public warnings and for study by researchers.

Formation and Dissipation of Hurricanes

The birthplaces of tropical storms are found over the oceans about 5 to 15 degrees of latitude from the equator. There is still some uncertainty as to the exact mechanism by which they start, how they intensify and why they follow the observed paths, but over the last ten years or so a great deal has been learned about all these points.

The sequence of events that must occur for the production of a hurricane has been given by H. Riehl*, a recognized authority on the subject. There must be a source of energy to start and maintain the vortex. As in most atmospheric vortices, the latent heat realized by the condensation

* H. Riehl, *Tropical Meteorology,* McGraw-Hill Book Co., 1954.

of water vapor is the major form of energy (see Chapter 1).

Since vertical air motion is needed to bring about condensation, which in turn leads to the latent heat release, there must be a suitable wind arrangement to act as a starting mechanism. Once upward air motion has begun, it would be accompanied by inflow in the lower levels and outflow in the upper levels of the atmosphere. Because of the influence of the earth's rotation the converging air will be turned and start to move in a circular path. This is the Coriolis effect you read about in Chapter 3. In the Northern Hemisphere the air moves in a counterclockwise circle. In the Southern Hemisphere the reverse happens. As the air moves closer to the center of the vortex, it will spin faster and faster.

A consideration of these various factors has led Riehl to propose a new theory on hurricane formation. He has observed that most hurricanes form over those ocean areas where, and in those seasons of the year when, the sea surface temperature is highest. In these circumstances the supplies of latent heat as well as the so-called sensible heat reach maximum values. The sensible heat is that transferred from a warm body to a cool one; the amount transferred is related to the temperature difference of the bodies. The addition of heat to the lower layers of the atmosphere increases the air temperature and tends to make an air mass unstable, and therefore, promotes convection.

In order for organized vertical motions to begin, there must be a coincidental arrival, over the warm ocean, of a low level and a high level *disturbance* in the general wind field. It is known that in the latitudes where hurricanes usually form the trade winds prevail. In the general steady current at the lower altitudes there occur frequent disturbances, that is, regions in which the wind changes direction in such a way as to produce widespread upward air motion and a lowering of the surface pressure. These conditions are accompanied by convergence of air into the disturbances or *waves,* as they are commonly called.

In the higher levels of the atmosphere there may also occur waves of such a type that over distances of the order of 100 miles the air tends to converge or diverge. Conditions for hurricane formation are ripe when a diverging system at high levels moves over a converging system at low levels. Then the circulation of air will be suitable for the development of an intense vortex. As the low-level air converges toward a certain point, it rises because the earth's surface restrains it from going downward. The high temperatures of the air resulting from heat conduction from the ocean surface also promote upward air motion. Once ascent begins, heat released during condensation increases the air buoyancy and produces upward acceleration of the air. As the air approaches higher altitudes, the diverging wind pattern caused by the upper level disturbance supplies a ready mechanism for removing the rising air to distant regions where slow descent occurs.

As long as the temperature in the region of rising air is higher than that outside it, the upward motion will continue to increase in magnitude. This will be accompanied by a decrease in surface pressure at the center of the column. As the pressure difference between the center of the incipient vortex and the surroundings increases, there is an accompanying increase of the speed of the converging air and of the velocity of the air moving around the center of the vortex. As long as the supply of energy can overcome forces tending to exert a braking action, such as friction, the intensity of the vortex will continue increasing.

The sequence of events described in the preceding paragraphs may take anywhere from as little as 12 hours to several days. Through the formative stage the winds usually do not reach hurricane force, being in the vicinity of 40 miles per hour, and are usually strongest in the quadrant ahead of, and to the right of, the center of the moving vortex. In this phase of the hurricane's life, the pressure in the storm center shows a gradual decrease. As the storm approaches maximum intensity, changes occur with greater rapidity. The pressure falls rapidly, winds increase to over 100 miles per hour in a narrow band 20 to 30 miles in diameter around the center of the hurricane, and the clouds and rain become organized into spiraling bands around the storm. When the hurricane reaches maturity, the pressure decreases at the center are relatively small, but the area covered by strong winds and rain increases. The region affected by hurricane winds may increase to a diameter exceeding 200 miles.

Hurricanes begin to weaken and die when the source of energy is diminished, usually when the storm moves over land. At one time it was speculated that hurricane dissipation was caused by the increase of friction of the land mass, that is, that physical bodies such as trees, mountains, etc., exerted forces on the wind that would have the effect of slowing down the vortex. It is now known that this explanation is not the whole story by any means. The major reason for the demise of a hurricane as it moves over land is that the supply of warm, moist air is reduced. This reduction in turn reduces the supply of available latent heat of condensation. Similarly the movement of a hurricane to higher latitudes over colder ocean water also leads to the dissipation of the hurricane because the lower the temperature, the smaller the quantity of water vapor and the smaller the supply of latent heat.

Hurricane Eye

One of the most interesting aspects of a hurricane is the eye. Over the centuries it has probably been the cause of many disasters. It is easy to believe that the storm has passed when the rain and strong winds stop. An assumption that this is actually the case could have tragic results when the rear part of the hurricane suddenly strikes. I. R. Tannehill* has cited the following colorful account by the Rev. J. J. Williams, S.J., of a hurricane which struck Black River, Jamaica, in 1912:

"Then succeeded a breathless calm for a few hours, that seemed to indicate that the very vortex of the storm was passing over us. This lull lasted for three hours. The unnatural stillness, marred only by an occasional drizzle, was itself portentous of approaching trouble. As there had been no change of the

* I. R. Tannehill, *Hurricanes,* Princeton University Press, 1952.

wind, the knowing ones prepared for the worst. . . . The rain was coming in fitful gusts, when suddenly we seemed to be standing in the midst of a blazing furnace. Around the entire horizon was a ring of blood-red fire, shading away to a brilliant amber at the zenith. The sky, in fact (it was near the hour of sunset), formed one great fiery dome of reddish light that shone through the descending rain. . . . Then burst forth the hurricane afresh, and for two hours or more (I have lost track of the hours that night) it raged and tore asunder what little had passed unscathed through the previous blow."

It has been found that hurricane eyes average about 15 miles in diameter but may reach 40 miles in very large storms. Winds are usually below 15 miles per hour and sometimes are at a dead calm. Cloud conditions in the eye may vary over a wide range. At times there may be only a few clouds, but usually there are many with interspersed open patches through which the sky can be seen. This is in sharp contrast with the outer edge of the eye, where a very deep layer of cloud extends from near the ground to great heights. It is common to describe the edge of the eye as the *cloud wall*. An unusual photograph of the eye of a hurricane is shown in Plate IX. The breaks in the clouds are evident as is the solid wall of clouds in the background. However, the swirling white mass is not commonly observed.

Over the oceans, the eye of a hurricane does not offer respite to a ship as it does to a land-lubber. The intense winds surrounding it create frighteningly huge waves. The following citation is from John Elliot's *Handbook of Cyclonic Storms of the Bay of Bengal.** It relates the experience of the ship *Idaho* as it passed through the eye of a typhoon in the China Sea on September 21, 1869.

"Till then the sea had been beaten down by the wind, and only boarded the vessel when she became completely unmanageable; but now the waters relieved of all restraint (in the calm center) rose in their own might. Ghastly gleams of lightning revealed them piled up on every side in rough pyramidal masses, mountain high—the revolving circle of the wind, which everywhere enclosed them, causing them to boil and tumble, as though they were being stirred in some mighty cauldron. The ship, no longer blown over on her side, rolled and pitched, and was tossed like a cork. The sea rose, toppled over, and fell with crushing force upon her decks. Once she shipped immense bodies of water over both her bows, both quarters, and the starboard gangway at the same moment. Her seams opened fore and aft. Both above and below the men were pitched about the decks and many of them injured."

From the two eyewitness accounts given here, it is quite evident the conditions during the passage of a hurricane are such that completely objective observations are difficult. It is reasonable to conclude that reports have to be accepted with a certain amount of caution. The fact that an observer in the eye has just experienced several hours of

* From Tannehill, op. cit.

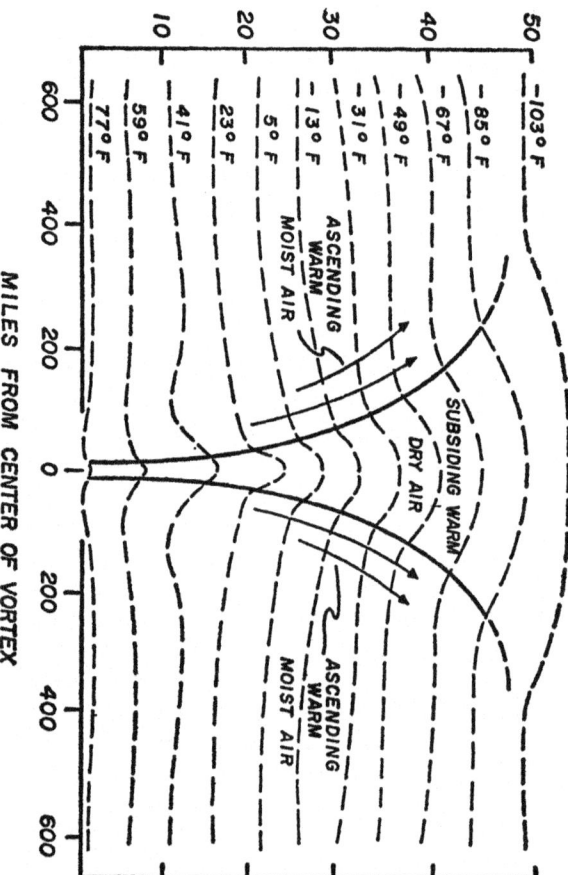

FIG. 15. *Temperature distribution in a mature hurricane is charted in this model, with the solid lines denoting the eye boundary. From E. Palmen, Geophysica (Helsinki), Vol. 3, 1948.*

strong winds and heavy rain prior to the arrival of the calm eye may explain the reports that conditions are hot and sultry. With a few striking exceptions, temperature measurements at the surface do not show that the eye is significantly warmer than the surrounding region. However, at higher levels the inner core of the hurricane becomes substantially warmer than the surrounding air. As you can see in Fig. 15, near the earth's surface the temperature was fairly uniform across the vortex, but it decreased more rapidly with altitude outside the core. For example, at 20,000 feet, the air at the edge of the core was 22° F, while at the center of the core it was about 32° F, or about 10° F warmer. At a distance of 400 miles from the vortex center, the temperature was about 11° F.

The observed temperature distribution must be related to the pattern of vertical air motion. It has been found that most of the upward air motion is restricted to the zone just outside the eye where warm, moist air prevails. The buoyancy of this air is greater than that at the outer reaches of the vortex, and therefore it can be expected to rise. The vertical variation of temperature of the ascending air is close to what one would expect if the air had risen from the ocean surface according to the adiabatic laws.*

For the high temperatures observed in the eye of the hurricane it is necessary that there be downward motion of cloudless air. Such subsiding air would warm up dry adiabatically. A satisfactory explanation has not yet been given of why the air in the eye descends in such a manner as to account for the observed temperatures.

* See pages 30 to 32.

Rainfall Patterns in Hurricanes

It has long been known from rainfall measurements made with widely spaced rain gages that in hurricanes affecting the eastern United States, most rain falls in the right-forward quadrant. To the left and rear of the storm center the weather may be relatively tame. Unfortunately, from rain gage records alone it was not possible to obtain a good picture of the pattern of clouds and rain. Since the middle 1940s we have learned a great deal about these things because of the introduction of radar and the institution of regular programs of air reconnaissance of hurricanes.

One of the most striking things found by radar observations is that once a hurricane has formed, the precipitation becomes surprisingly well organized. The showers and thunderstorms, rather than being scattered in a random fashion, usually are lined up as spiral bands which converge toward the center of the hurricane (Plate X). The bands are made up of showers and thunderstorms, which form and dissipate rapidly but are replaced by new ones along the same bands. Near the centers of hurricanes circular bands sometimes replace the spirals.

Radar observations have shown that 300 to 400 miles ahead of the hurricane center, one frequently finds well-defined fairly straight lines of thunderstorms which move in the direction of the storm.

On at least one occasion circular or spiral cloud bands have also been seen in the eye of a hurricane (Plate XI).

A completely acceptable explanation for the

bands of clouds and rain has not yet been offered, but it has been suggested that they develop this banded structure for the same reason that lines of convective clouds occur in less disturbed atmospheric systems.* It is speculated that the organization results because the clouds align themselves roughly in the direction of the winds.

Movement of Hurricanes

A consistent property of all tropical storms is that, once formed, they follow paths that carry them poleward. In general Atlantic hurricanes have only a small poleward component initially, but after perhaps a few days' movement along a nearly east-west path, they start curving toward the north. This is clearly illustrated in Fig. 16, taken from Tannehill, which plots the paths of hurricanes observed in the first two weeks of August during the 59-year period from 1874 to 1933. It can be seen that occasionally hurricanes may deviate from smoothly curved paths. They sometimes undergo rapid changes of direction, even going so far as to loop around a particular area. When this occurs, the duration of the storm over that area can be considerably longer than normally would have been expected, and the inhabitants pay for nature's abnormality by suffering heavy rains and strong winds for very long periods.

Young hurricanes in the tropics move fairly slowly, averaging 12 to 15 miles per hour. As the storms intensify and start recurving, their speeds increase. At times Atlantic hurricanes may reach

* See pages 37 to 38.

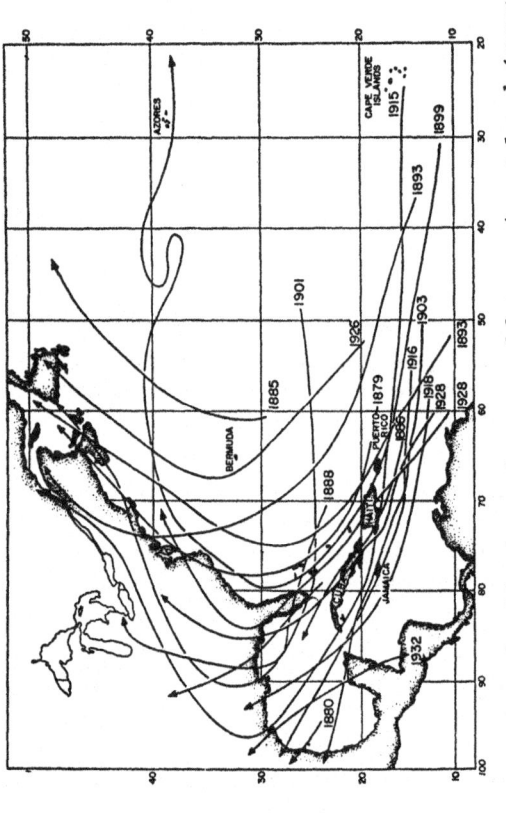

FIG. 16. The paths of hurricanes that occurred between August 1 and August 15 in the years from 1874 to 1933 are plotted on this chart. From I. R. Tannehill.

50 to 60 miles per hour as they impinge on the northeastern coast of the United States.

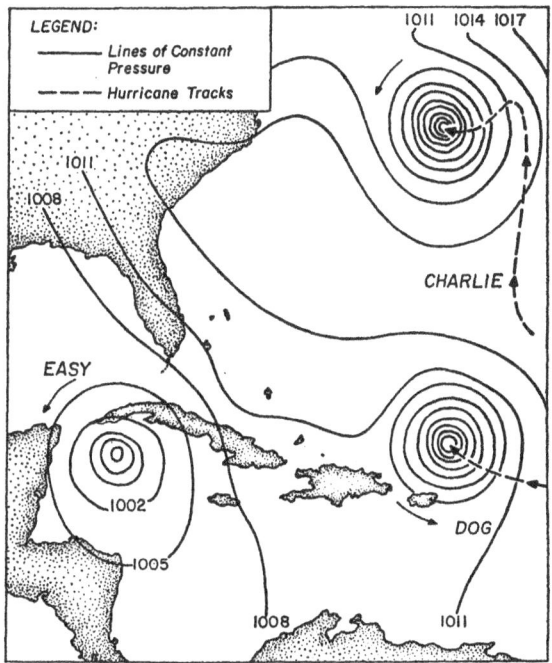

FIG. 17.   *Three hurricane vortices appear on this weather map for 7:30 A.M. on September 2, 1950. Broken lines show the pre-map time tracks of Hurricanes Charlie and Dog. From G. Norton,* Monthly Weather Review, *Vol. 79, 1951.*

The tracks plotted in Fig. 16 were obtained by following the hurricanes on weather charts showing the patterns of atmospheric pressure. Because of the low pressures in hurricanes, they stand out clearly. This is evident from the unusual map for

07:30 A.M. September 2, 1950 (Fig. 17).* The light solid lines represent lines of equal pressure in millibars (one inch of mercury = 33.9 millibars). Since, as shown in Chapter 3, the low-level winds blow almost parallel to these isobars, a map of surface pressure also gives information about the wind pattern. The air motion in these hurricanes was in a counterclockwise direction. Each of the circular low pressure areas represented a rapidly spinning vortex about 300 miles in diameter.

This situation, in early September 1950, is unusual, if not unique, in that three hurricanes were present at the same time in the vicinity of the United States. Most often one vortex dissipates before the second develops; occasionally a second forms while one is still in existence. The presence of three well-developed vortices at the same time is quite uncommon. These hurricanes, Charlie, Dog, and Easy, followed different tracks and clearly demonstrate the dangers of deciding how a particular hurricane will move from a picture of the average storm movements.

Hurricane Charlie spent its entire life far out in the Atlantic and never posed a serious threat to land areas. An airplane in the storm recorded winds to 115 miles per hour. It is interesting to note that the hurricane's track had a peculiar jog to the west and then a slow turn to the northwest. During the next two days the hurricane continued recurving until it was moving toward the northeast.

* The map shown here and the relevant discussion were taken from G. Norton, *Monthly Weather Review*, U. S. Weather Bureau, Washington, D.C., Vol. 79, Jan. 1951, pp. 9–12.

Hurricane Dog was the most severe of the 1950 season. The crews of aircraft flown into the storm estimated winds at over 180 miles per hour and waves 100 feet high. As Dog moved close to Antigua and Barbuda in the Caribbean, winds reached a maximum value of over 130 miles per hour and remained over 75 miles per hour for six hours. Damage was very heavy because of the strong winds, waves, and flooding. On September 12 the hurricane passed close to the New England coast leaving 12 people dead and $2,000,000 of property damage.

Hurricane Easy formed over the northwestern Caribbean Sea. For almost two days it sat over its birthplace and built up momentum before moving northward over Cuba and then northwestward parallel to the west coast of Florida about 30 to 50 miles offshore. When it was about 70 miles north of Tampa, it made a loop, which carried it close to the coast just south of Cedar Keys on the morning of the fifth. At this point it made still another loop. As a result of the looping movement of the hurricane center, the town of Cedar Keys was in the eye for two and a half hours and was exposed to the same side of the hurricane twice. Winds in the hurricane reached 125 miles per hour and were over 75 miles per hour at Cedar Keys for more than nine hours. During the three days when the town was under the influence of the hurricane, more than 24 inches of rain fell. Half the houses in the town of about one thousand people were demolished, and 90 per cent of the rest were damaged. All the fishing boats which served as the main source of income for the inhabitants were destroyed.

After Hurricane Easy left Cedar Keys, it moved southward about 70 miles and then made sudden turns to the east and then north. Fortunately, by this time it was beginning to weaken rapidly, and it dissipated as it moved into Georgia on September 7.

For reasons given earlier, when hurricanes move over large land areas, or over colder ocean water, they weaken rapidly. On a weather map, you see this as an increase in the pressure at the center of the vortex. Meteorologists refer to this process as *filling* of the low-pressure area. As filling occurs, the wind speeds decrease and the storm becomes a so-called *extratropical cyclone,* that is, a cyclone having properties common to cyclones at latitudes outside the tropics. The next chapter will discuss the properties of such storms in some detail. Here it suffices to say that they are large, roughly circular, low pressure areas, around which the winds blow with low and moderate velocities.

Hurricane Forecasting

When a meteorologist draws a weather map and detects an incipient hurricane, one of his first jobs is to locate the center, check its velocity and then predict its future track. The precise location of the hurricane is not as easily identified as one would think from examining the map shown in Fig. 18. If one had very detailed measurements of atmospheric pressure, the point of lowest pressure could easily be identified. Unfortunately, pressure measurements are rather sparse over land, and over the ocean they are widely separated. The distances between wind measuring stations are equally large, being of the order of 100 miles. As a result, the possible error in locating a hurricane center on the basis of pressure and wind observations may be about 50 miles.

At the present time the weather services utilize

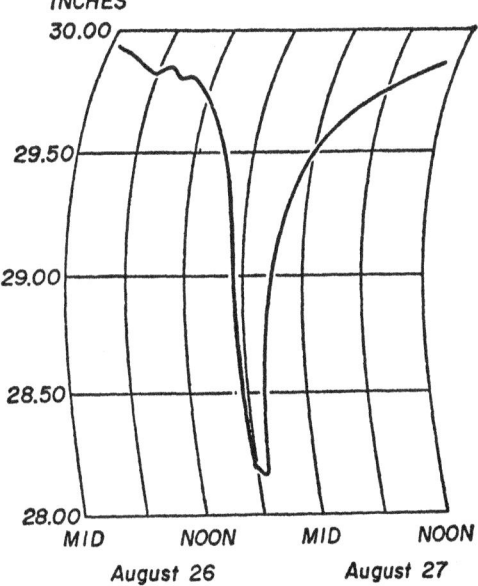

FIG. 18.   *The similarity and the difference in the pressure changes in hurricanes and tornadoes can be seen by comparing this hurricane pressure trace with Fig. 14. Note that the hurricane pressure drop here occurred over a period of hours, not of minutes as in Fig. 14. This trace was recorded during passage of a hurricane at West Palm Beach, Florida, on August 26, 1949. From R. T. Zoch,* Monthly Weather Review, *Vol. 77, 1949.*

much more information than is obtained at the typical weather station. Since late in World War II Air Force and Navy weather reconnaissance aircraft have probed regions where hurricanes are suspected. The U. S. Weather Bureau National Hurricane Project also has airplanes equipped to make special measurements in tropical storms. They locate the hurricane and fly around it, and if flight conditions permit, penetrate into the eye and obtain its precise location. With modern navigational equipment the aircraft position can be determined with an accuracy of perhaps a mile or two. Airplanes continue patrolling the storm until it strikes land or moves into the open oceanic areas. When a hurricane moves close to land, it can be tracked with good accuracy with shore-based radar sets. Present equipment can detect hurricanes out to more than 100 miles, and at times the upper parts of the thunderstorms in hurricanes can be detected at distances up to 200 miles.

With all the available information it is possible now to keep a careful watch on areas where hurricanes may be breeding, to spot them when they form, and to track them accurately. This is particularly true of hurricanes forming in the Western Atlantic Ocean, Caribbean Sea, and the Gulf of Mexico. The chances of a surprise hurricane striking the eastern half of the United States are small indeed.

Since present-day techniques permit us to locate a hurricane accurately, the big problem facing the meteorologist is that of forecasting future locations so that threatened areas can have adequate warnings. Available statistics show that over the last

25 years better observations, forecasts and warning systems have led to the saving of many lives. This is shown in the following table taken from U. S. Weather Bureau records.

Table 2

Loss of life and property caused by hurricanes and the average loss of life per $10,000,000 of property damage.

PERIOD	TOTAL PROPERTY DAMAGE	TOTAL LOSS OF LIFE	AVERAGE LOSS OF LIFE PER $10,000,000 OF PROPERTY DAMAGE
1926–1930	$131,153,000	2108	161
1931–1935	60,910,000	494	81
1936–1940	257,333,000	663	26
1941–1945	296,924,000	107	4
1946–1950	253,700,000	69	2.7

As far as property damage was concerned there was no significant decrease over the period 1926 to 1950. But during this interval there were extensive property developments in areas such as Florida so that the damage potential increased rapidly. Where once there were swamps, luxury hotels and expensive homes were built. Thus, a hurricane in the late forties had much more to knock down and inundate than did a hurricane in the early thirties. In addition, inflation had its effect in increasing dollar damage.

Notwithstanding the difficulties in interpreting the property-damage statistics, the figures on loss of life show an impressive decrease over the period. This is particularly true when the fatalities

are measured in terms of damage (last column).
The use of this scheme of measuring fatalities has
some validity because the property damage is, at
least to a certain extent, a measure of the number
of inhabitants likely to have been affected. The
June 1957 hurricane mentioned at the beginning
of this chapter will cause the extension of this table to show a rise in the number of deaths during
the 1955 to 1960 period, but, nevertheless, it is
evident that the number of lives lost has fallen far
below what it would have been 25 to 30 years
ago.

Various schemes are used by meteorologists to
predict the future course of a hurricane. Once the
past positions are accurately known, a first es-
timation as to where it will be in the future can
be obtained by extrapolation. As shown by the
map in Fig. 17, hurricanes often follow a regular
track. Often the assumption that the past behavior
will govern future behavior, at least for a period
of a few hours, is quite good. Unfortunately, such
a simple extrapolation procedure is of questionable value for long time periods and tells noth-
ing about erratic changes of direction.

Modern hurricane forecasting techniques recognize that a hurricane is a spinning vortex em-
bedded in a larger mass of air whose motion is
largely governed by forces outside the hurricane.
To a great extent the hurricane is carried along by
the large-scale motion. In addition, the air in the
vortex leads to forces tending to make it move
poleward, but this factor is of secondary importance. The forecaster's main task is to predict the
circulation of the large body of air encompassing
the hurricane. For periods of about 24 hours fairly

accurate forecasts can be obtained by considering the average motion of the air within about 1000 miles of the storm.

For making forecasts two to three days in advance, it is necessary to consider the airflow over most of the Northern Hemisphere (when the hurricane is in this hemisphere). The airflow in the eastern United States, for example, is part of the huge mass of air flowing in a relatively thin shell over the hemisphere. Disturbances in the flow many thousands of miles away can propagate as waves more rapidly than the wind, and be felt in the eastern United States in shorter periods of time than if the effects were carried with the wind. On the basis of the airflow over the hemisphere, the forecaster must first predict what the flow pattern will be within about 1000 miles of the hurricane, and then use this information to predict future hurricane positions. In practice it has been found that the airflow at about 20,000 feet is a good measure of the general airflow below the stratosphere, and winds at this altitude are used extensively for hurricane forecasting. This level is often called the *steering level* because, loosely speaking, the wind flow here can be considered to "steer" the hurricane.

At the present time the accuracy of a forecast of the hurricane center issued one day in advance is considered excellent if the predicted position of the center is within 50 miles of the correct position. It should be recognized that when issuing a 24-hour forecast, weather data 30 to 36 hours in advance must be used. Errors exceeding 50 miles are not surprising when one considers the factors involved. If a hurricane is moving at 15 miles per hour, the error in placing the center 36 hours in advance will exceed 50 miles if the predicted course is off only 10° on the 360° compass. At higher speeds the error, of course, will be correspondingly larger. From simple calculations such as this one, it is evident why very accurate predictions of hurricane velocities are needed in order for them to be useful to the public. In view of the uncertainties of the forecasted hurricane position and the variability of the size of the area of hurricane winds, the warning area is usually substantially larger than the one that actually will sustain damage. The meteorologist chooses to err on the safe side. When human life is at stake, hedging the prediction certainly is the prudent course of action.

Ocean Waves Produced by Hurricanes

The strong winds in a hurricane exert forces on the ocean surface and generate huge waves which propagate outward in all directions. In some cases they can be detected as long sea swells far away from the vortex. The waves created in the rear-right quadrant of the storm move in the direction of the storm. These waves are the strongest produced by a hurricane and may have a speed of propagation of 1000 miles per day. Since hurricanes move from 300 to 400 miles per day, the arrival of strong sea swells may be indicative of the presence of an approaching hurricane some 600 or 700 miles out to sea.

Reliable observations of the ocean surface behavior in the vicinity of a hurricane do not exist. It appears that along a coast line there is a "piling up" of water because of the prolonged periods of strong winds. This effect may cause a rapid increase of the water level by as much as 10 feet above the normal tide. Superimposed on the high water level are waves which propagate rapidly. Reports of a "wall of water" suggest that a single wave of tremendous proportions is created by the storm. In some hurricanes many large waves are observed.

Hurricane Damage and Safety Measures

During the period 1926 to 1950 hurricanes caused about $1,000,000,000 worth of property damage and caused almost 3500 deaths (see Table 2, page 119). Since 1950 there have been perhaps another 1000 fatalities and $500,000,000 of property damages.

The most vicious hurricane on record from the point of view of lives lost was the one known as San Felipe, which moved through the Caribbean on September 13 and 14, 1928. As it passed over Puerto Rico, 300 people were killed. During passage over Florida, it caused 1836 deaths.

Most fatalities and property damage occur in coastal areas. There are various reasons why this is so. When hurricanes move over land, they tend to weaken rather quickly. Wind velocities and rain intensities decrease. Indeed, people who live far inland sometimes find that hurricanes are beneficial by bringing heavy, but not torrential, rains for increasing ground moisture and replenishing water supplies. But when hurricanes pass over the coast line, they may be at maximum intensity. The winds can be blowing at speeds high enough to blow over houses, rip down high-tension power lines, and carry away boats, automobiles and anything else which is not very heavy or securely lashed down.

The second factor which must concern all people living on low-lying land near the coast lines is the effect of the hurricane in causing inundations by the sea. Also, in mountainous areas the torrential rains can produce flash floods of terrifying proportions. It is not uncommon for a hurricane to dump more than 20 inches of rain during its passage over a particular point. As much as this has been recorded in a single day. Riehl* has reported that a typhoon over the Philippines produced an extreme of 100 inches. For comparison, this is two to four times the average precipitation falling during the entire year at most stations in the United States. The major losses of life and property can be ascribed to drowning and flooding following "tidal" waves and torrential rains. The only safe course of action for residents of coastal lands threatened by a hurricane is evacuation to higher ground.

In the chapter on tornadoes we pointed out that they cause destruction because buildings explode in the sudden lowering of pressure as the funnel passes. One of the striking features of hurricanes is that the pressure is quite low at the center. As Fig. 18 shows, the trace of pressure change pictorially looks somewhat similar to that in a tornado, but there is a very important difference. Although the pressure in a hurricane may fall to values as low as those observed in some tornadoes, the drop of pressure occurs over a period of the order of six hours or more rather than the minute or so in a tornado. During the many hours of gradual

* Riehl, op. cit.

pressure drop as the hurricane moves over a building, the pressure on the inside can accommodate itself to the falling pressure outside though the windows and doors are closed. Even small openings usually will allow enough air movement to prevent the pressure inside the house to become dangerously greater than that outside.

Since there is usually no "explosion threat," windows may be closed and shuttered or boarded up against a hurricane. However, to be on the safe side, a window on the leeward side of the house can be opened to accelerate the pressure adjustments. Winds over 100 miles per hour can tear down many buildings and carry the remnants through the air like shrapnel. Covering of windows prevents the debris from flying through them.

In some hurricanes tornadoes develop along the lines of thunderstorms. When this occurs, the inhabitants of the affected area are subjected to the worst possible weather dangers.

The following report is an illustration of the combined lethal powers of tornadoes and hurricanes. It is taken from an article by W. R. Davis* which describes the history of Hurricane Gracie. In late September 1959, Gracie caused $14,000,-000 damage and took a total of 22 lives.

"Ten persons lost their lives in South Carolina and Georgia, mostly due to automobile accidents, falling trees, and live wires. Several tornadoes attended the passage of the dying storm through Virginia and twelve persons were killed in one of these tornadoes at Ivy near Charlottesville."

* W. R. Davis, *Weatherwise*, American Meteorological Society, Vol. 13, Feb. 1960, p. 24.

The occurrence of tornadoes probably is more frequent than once was thought. For obvious reasons, conditions during the passage of a hurricane are such as to make it quite difficult to obtain good observations. Also, after the hurricane has passed, the destruction which may have been caused by a narrow, intense, short-lived tornado vortex cannot easily be distinguished from the effects of the hurricane itself.

On the basis of many years of experience, the U. S. Weather Bureau has formulated a set of rules to follow in order to increase chances of survival when a hurricane threatens.

1. Keep your radio or television set turned on and listen for the latest official Weather Bureau advisories.
2. Before the strong winds and floods strike, be sure your car is ready to evacuate you and your family. Fill the gas tank and check the battery and tires.
3. Have a supply of drinking water and food on hand. Select foods which do not need cooking or refrigeration. Have a flashlight, first-aid kit and battery-operated radio on hand. Remember that when the storm hits, electric power will probably be knocked out for many hours or possibly days.
4. Store all loose objects such as toys, tools, trash cans, etc. Board up all windows. A window on the side of the house away from the storm can be left partly open.
5. Get away from low areas that may be swept by storm tides or floods before the storm arrives.

6. During the storm, stay indoors. Do not be fooled if the calm eye passes over. Keep your radio or television set on and listen for information from the Weather Bureau, Civil Defense, Red Cross, and other authorities.

7. After the storm, proceed out of your house with caution. Do not drive unless it is necessary to do so. Watch for undermined pavement and broken power lines. Be careful to prevent outbreaks of fire because of broken oil or gas lines. Report downed power wires, broken water or sewer pipes to proper authorities. Use your telephone for emergencies only.

This set of instructions sounds like procedures to be followed during a bombing raid. Unless proper safety measures are followed, the consequences of a hurricane can be the same—injury and death.

CHAPTER 7

CYCLONES

What would you do if a weatherman announced, "A cyclone is coming"? Some people would run for cover. However, they would probably be running unnecessarily. In its most general sense the term *cyclone* refers to an area of low pressure around which the air is circulating. In the Northern Hemisphere, cyclonic motion is counterclockwise; in the Southern Hemisphere it is the reverse. From this definition you can see that a hurricane is one kind of cyclone, a tropical cyclone.

When meteorologists speak of cyclones, they usually are referring to those nearly circular areas of low pressure which form outside the tropics. The term *extratropical cyclone* is used when one wants to be sure of avoiding confusion.

If you look at the weather maps printed in many newspapers, one or more cyclones almost always can be seen. Television weathermen refer to them as centers of low pressure. Prolonged periods of rain and snow in the winter are caused by their arrival. They are the largest storm centers nature produces but, fortunately, are also the tamest. The winds usually are not strong enough to cause damage to property, injuries, or fatalities. On the other hand, they bring rain and snow for filling reservoirs, increasing soil moisture and permitting plants to grow.

Sometimes, but fortunately not very often, cyclones lead to the fall of rains heavy enough to cause floods. The snows and blizzards of the winter months are brought by the passage of cyclones. In some years families of cyclones move over particular areas and cause the accumulation of huge quantities of snow which isolates towns and cuts off food supplies for livestock. When this happens, a first-class disaster may be in the making. Since the damage and deaths are not of the sudden violent type caused by tornadoes and hurricanes, they somehow do not seem to be quite as dramatic. But they are just as tragic.

In some years the worst effects of cyclones are felt weeks or months after their passage. The spring floods which plague the Midwest as the Ohio, Missouri, and Mississippi Rivers rise above flood stage fall in this category. Heavy snows may accumulate on the ground over a period of weeks with little change unless the temperature rises high enough to melt them. With the advent of spring the storm track moves northward, and the stage is set for trouble. Cyclones passing over the snow-covered area in these circumstances could bring rain and high temperatures. This combination sometimes leads to very rapid melting and runoff into the streams and rivers. The results are swollen rivers and floods with all the terrible consequences.

Cyclones are the largest *storms,* but the largest vortices in the atmosphere are the *anticyclones.*

The name is a natural one; anticyclones are regions of high pressure in which the winds blow clockwise, that is, in the opposite direction to a cyclone's. Anticyclones may be a few thousands of miles in diameter and persist for very long times. The winds blowing around them are generally light and the skies are free of rain clouds. Measurements show that in high-pressure areas the air usually sinks and in the process it dries. This discourages the formation of clouds and leads to fair weather. So, when a barometer shows that the pressure is rising, you often can assume that an anticyclone is approaching and that fair weather is on the way.

But our primary interest in this book is in the storms, and we shall pass by the anticyclones with only this nod of friendly recognition.

Difference Between Cyclones, Hurricanes, and Tornadoes

The cyclones that commonly form in middle latitudes and slowly move across the country differ in several important respects from the other vortices discussed in earlier chapters. To someone who has had the misfortune to experience all types of storms, one of the biggest differences among them would be in the strength of the wind. In a tornado winds may exceed 300 or 400 miles per hour, in a hurricane they may be above 150 miles per hour, but a typical cyclone usually has winds below 50 miles per hour. Fortunately, the durations and sizes of the storms vary in the reverse order. A tornado usually is less than a mile wide and lasts for a few minutes; a hurricane may be 500 miles in diameter and last for as much as a week or so; a cyclone may be more than a 1000 miles in diameter and last for perhaps a week. Evidently a hurricane much more closely resembles the typical cyclone than does a tornado. As a matter of fact, when a hurricane moves over land or cold ocean water, it slows down and increases in size and evolves into a common extratropical cyclone.

Formation of Cyclones

There is still a certain amount of disagreement among meteorologists about the main factors in cyclone formation. It is evident from a study of weather charts that when a low-pressure area develops and intensifies, it extends to high levels of the atmosphere, up to perhaps 30,000 feet. Since these storms form in the wintertime, when the stratosphere is low, the storm system extends as high as the stratosphere. One of the uncertainties about cyclone generation is centered on the question of where is the generation initiated, at the low levels or high levels, or does there have to be a superposition of a high and low level disturbance before an intense vortex will form?

The first comprehensive theory on cyclone development was advanced in 1918 by a group of Scandinavian meteorologists, led by J. Bjerknes at the Geophysical Institute in Bergen, Norway. It is called the *polar-front theory* of cyclones. For many years it has been accepted as satisfactory because it explains many of the observed properties of cyclones. However, the theory was developed at a time when there were few observations of wind, pressure, temperature, and humidity in the higher levels of the atmosphere. As they have become more numerous, certain aspects of the theory have been questioned. Nevertheless, it still represents an important step in our understanding of the ways of the weather and is used extensively by forecasters.

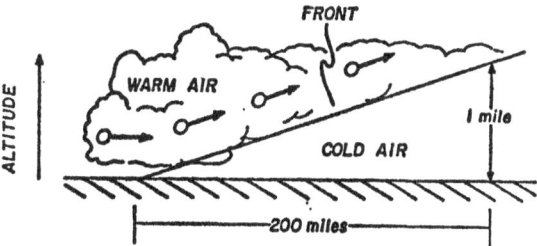

FIG. 19. *At a warm front cold air slides under warm air and causes the warm air to rise.*

The Scandinavian meteorologists were the first to recognize the importance of the fact that when a cold mass of air meets a warm mass of air, the two do not mix rapidly. Instead, the cold air slides under the warm air and causes the latter to rise (Fig. 19). The boundary between the two air masses is rather sharp and has been called a *front*. Since the frontal theory was developed during the First World War, the boundary between two differing air masses was likened to the front separating opposing armies.

When cold air advances along the surface of the earth and replaces warm air, the boundary is called a *cold front*. When the cold air retreats, the boundary is called a *warm front*. If you examine weather maps of the Northern Hemisphere, you find that a common feature is an extensive front separating air of polar origin from air of tropical

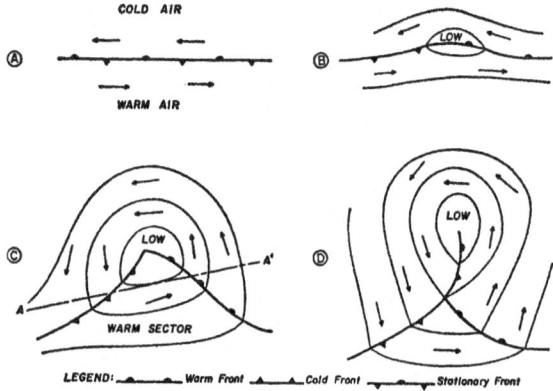

FIG. 20. *A cyclone may develop as a wave on a polar front. At* (A) *air on the two sides of the front may be moving in opposite directions. At* (B) *perturbation in the wind flow generates a small wave. At* (C) *the amplitude of the wave increases, circular motion begins and warm air moves up the frontal surface. At* (D) *fronts have occluded as the cold front advanced under the warm front.*

origin. This narrow transition zone from cold to warm air is called the *polar front*. It may extend for thousands of miles and last for weeks.

On the northern side of the polar front, the air may be moving from the east, while to the south it may be moving from the west (Fig. 20). In this case there is little upward air motion. Perturbations in the wind flow often cause bulges in the frontal surface, perhaps because of the deflection of the air by a mountain range or the temperature difference between land and sea. A small *wave* develops in the front. Under certain conditions the amplitude of the wave increases. This does not always happen. Sometimes the front straightens out and the wave disappears. At other times, when the wind, temperature, and humidity distributions lead to instability, the wave amplitude increases rapidly and circular motion of the air begins. At the same time warm air begins to move up the frontal surface. Condensation starts and the pressure at the center of the wave begins to fall. Then the air in the lower altitudes converges toward the center. As we saw in Chapter 3 the influence of the earth's rotation causes this air to move counterclockwise around the wave center. In a particularly unstable situation, wave development proceeds rapidly.

As can be seen in Fig. 20, the wave becomes composed of a warm front and cold front. Sometimes the latter moves more rapidly, and catches up to the warm front. With the passage of time the fronts become *occluded,* that is, the advancing cold front advances so far under the warm front that at the center of the vortex there may be only cold air near the ground. The formation of a well-developed wave may occur in a matter of hours;

the whole sequence shown in Fig. 20 may consume
a few days.

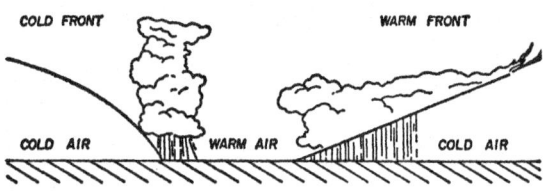

FIG. 21.   *A cross section of the weather situation
in Fig. 20 (C) would look like this along the line
AA'. At this point the cyclone is well developed.*

The early theories of wave formation on polar
fronts maintained that the cyclone formation depended mostly on the properties of the air masses
on opposite sides of the front. Nowadays it is accepted that the development of the storms is also
related to the air motions in the high levels of the
atmosphere. There is evidence to show that cyclonic centers form and intensify when particular
types of disturbance in the high level wind flow
move over the polar front. You will recall that the
same kind of superposition of high and low level
disturbances has been proposed as a requirement
for hurricane formation.

Until a cyclonic center begins to develop, the
polar front usually will not produce extensive
cloud systems. But when a wave begins to intensify
and vertical motions increase, clouds form over
large areas. The energy released in the condensation contributes to the vertical motion. It combines
with the energy contributed by the cold heavier air,
which tends to sink and spread outward. Over the
warm front there is slow but steady upglide of

moist air, and layer-type clouds form (Fig. 21), which may be seen hundreds of miles ahead of the cyclone center. As the center, moving from the west, approaches, the cloud bases are found at lower and lower altitudes, and more or less steady precipitation begins to fall. When the temperatures at the ground are below freezing, the precipitation is in the form of snow. If warmer temperatures prevail, the snow melts to give rain. In some cyclones the area of steady snow or rain can be a few hundred miles across. If the storm center moves slowly, large quantities of snow or rain may accumulate.

Once the warm front has passed, the observer will find himself in the warm air where widespread cloud layers normally do not develop. This area is known as the *warm sector*. A second band of precipitation is sometimes found along the leading edge of the cold front. The advancing wedge of cold air is steeper than the retreating warm front. Also, the air in the warm sector ahead of the cold front may be quite instable. Pronounced upward motions may occur, and showers and thunderstorms often form. The violent squall lines discussed in Chapter 4 most often develop in the warm sector ahead of, and nearly parallel to, the cold front.

Regions of Cyclone Formation and Cyclone Movements

Unlike hurricanes, which always form over warm oceans, middle latitude cyclones may form over land or sea. As a matter of fact, they may form virtually anywhere along the polar front. However, there are certain favored breeding grounds.

Over North America, cyclones frequently form just to the east of the Rocky Mountains. This fact has led many to the conclusion that the storms are initiated, at least in part, by the effects of the mountains. The average path taken by these vortices is slightly toward the east-southeast and then north of east. These are the storms that sweep across the central and northeastern United States and bring blizzards and cold weather.

Many cyclones form in the Gulf of Mexico and follow a track to the northeast along the eastern coast of the United States and then into the northern part of the Atlantic Ocean. Cyclones that strike northern Europe often are initiated over these same Atlantic waters. Storms affecting the Mediterranean region frequently come into being in the vicinity of Spain and move eastward.

Over the Pacific Ocean there are several favorite spawning areas. Many vortices develop over the ocean to the east and south of Japan and then tend to follow a northeast track. Cyclones often form over Siberia and move toward the east over the Islands of Japan and bring severe weather and cold waves.

Cyclones that strike the western coasts of North America form along the polar front running across the northern part of the Pacific Ocean. Many also begin over the ocean several thousand miles west of California. When these storms strike the west coast of the United States, they are often in an advanced stage of development. They bring rainfalls of major importance to the economy of the western states.

If you plot the tracks followed by cyclones, you see that on the average they move from southwest to northeast. There are at least two important factors governing the movements. The cyclones form as waves, which tend to move up the southwest-northeast oriented polar front. Such behavior is specified by the polar-front theory of cyclones. It has also been found that the cyclones tend to be "steered" by the prevailing winds at the middle and high altitudes. These winds on the average move from the west or southwest in the regions where cyclones are found.

It is well to keep in mind that most cyclones do not exactly follow the average paths mentioned in the preceding paragraphs. One major factor involved is the position of the polar front. In the winter it is farther south than in the summer, and the storms form at lower latitudes during the cold season of the year. Also, the "steering currents" usually deviate somewhat from the average pattern.

Individual cyclones differ from one another in many respects. Some have circular patterns of pressure, but most are elliptical with the longest axis in the north-south direction. Some may develop in a matter of hours; others may take days. Some may remain almost stationary for many hours, while others may move across the country at speeds of perhaps 40 or 50 miles per hour. On the average the speeds are 20 miles per hour in the summer and 30 miles per hour in the winter.

A single well-developed cyclone may dominate the weather for many days and then move off the continent, to be followed by an anticyclone and fair weather. Again, cyclones may form in rapid

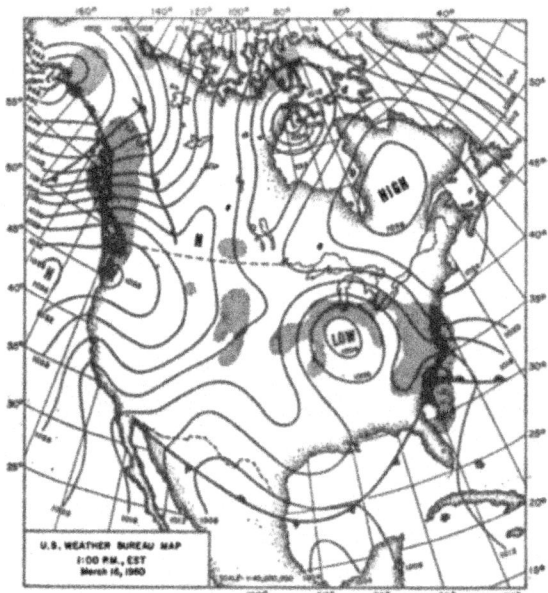

FIG. 22.   *The official U. S. Weather Bureau map
for 1:00 P.M. on March 16, 1960, shows several
cyclones. The shaded areas represent regions of
precipitation. The map shows rain along the east
and west coasts and snow over the central parts
of the continent. Twenty-four hours earlier the
cyclone shown over Illinois on the map was over
Oklahoma. A young wave is forming along the
Carolina coast.*

succession in the same area over a period of one or two weeks. Such a situation can lead to prolonged periods of bad weather broken by short periods of clear skies. The weather over most of the eastern half of the United States during March of 1960 was caused by this kind of situation. Cyclone after cyclone formed just east of the Rocky Mountains and moved toward the east and in the process dropped huge quantities of snow (Fig. 22). With each succeeding storm, the snow piled higher and created many records for the amount measured in one year.

The differences from one cyclone to another must be faced by the weather forecaster virtually every day. In order to make accurate predictions of temperature and rainfall he must predict the formation, intensification, and movement of cyclones. Until a few years ago this was done with forecasting rules that involved rather elementary types of calculations, and the accuracy of the forecasts depended to a large extent on the experience of the forecaster. Today, more and more use is being made of high-speed electronic computers, which in short periods of time, can make calculations from the complex equations describing the air motion over a large part of the Northern Hemisphere. This modern approach to atmospheric problems promises better understanding of the weather and, ultimately, better forecasts.

CHAPTER 8

CONCLUSION

In a single short book it is obviously impossible to dig very deeply into the intricate ways by which nature produces the storms which affect us all. However, we hope that the reader has gained a better understanding of the most important causes of winds and weather.

It should be recognized that there are many aspects of storm formation, movement, and dissipation that cannot yet be explained, but we should also note that explanations are coming at an increasing rate. Meteorology, or the physics of the atmosphere as it is appropriately called by many scientists, is rapidly being recognized as one of the most important branches of geophysics because it deals with the environment in which we live. As increasing attention is given to the problems, so will there be an increasing number of answers obtained.

We have seen that vortices in the atmosphere are produced when the pressure, temperature, humidity, and wind are distributed in such a fashion as to lead to instability and vertical air motion.

When there is sufficient energy available, great masses of air can be set into circulation. The basic laws of physics and thermodynamics are directly applicable in the study of these vortices. For example, the famous formula $F = ma$ (*force* equals *mass* times *acceleration*) applies to the atmosphere as it does to a stone dropped from the leaning tower of Pisa. The acceleration of a parcel of air increases in direct proportion to the force applied. The problem of the meteorologist is to specify properly the force or forces involved. As shown in Chapter 3, it is sometimes possible to do this and calculate wind speeds with acceptable accuracy. However, when one wishes to predict the formation of a hurricane, for example, from a mathematical model of the atmosphere, the problem becomes very complicated. Notwithstanding the difficulties, great progress is being made as a result of the combined efforts of meteorologists, physicists, and mathematicians. What we need are more imaginative, aggressive young people who are willing to accept the challenges presented by these difficult problems.

Another point worth noting is that progress in meteorology has been hindered by the lack of suitable observations. A cyclone may produce clouds and rain over an area hundreds of miles in diameter. To study the behavior of a cyclone it is desirable to have a detailed picture of the entire cloud system. When radar was introduced into meteorology some fifteen years ago, it became possible to examine the distribution of rainfall over a large part of a cyclone. But radar does not detect most clouds. In the past year we have obtained devices that can show the clouds over an entire cyclone. These devices are, of course, the Tiros I and Tiros II satellites, which transmit television photographs of the cloud structure over an area larger than an entire cyclone. Plans are already being made for weather satellites of improved design, which will give both cloud cover and radar observations of precipitation. Other satellites already have furnished information of the radiation from the earth to outer space and the incoming solar radiation. These new and dramatic sources of information will be of tremendous value in solving the many mysteries of the atmosphere.

We believe it is fair to say that with powerful new observational techniques, high-speed electronic computers, and the combined talents of mathematicians, physicists, and meteorologists, rays of knowledge are at last illuminating the clouds of ignorance.

ADDITIONAL READING MATERIAL

H. R. Byers and R. R. Braham, Jr., *The Thunderstorm,* U. S. Government Printing Office, 1949, 287 pp.

An authoritative book on all aspects of thunderstorms. Written at the completion of a research program of five years' duration. It is mostly descriptive and is suitable for intermediate and advanced students.

S. D. Flora, *Hailstorms of the United States,* University of Oklahoma Press, Norman, Okla., 1956, 201 pp.

A detailed description of all aspects of hailstorms. Extensive statistics are given of the properties of hail and its geographical distribution. It is written in popular terms.

S. D. Flora, *Tornadoes of the United States,* University of Oklahoma Press, Norman, Okla., 1953, 194 pp.

A detailed description of all aspects of tornadoes. Extensive statistics are given of the properties of tornadoes and their geographical distribution. It is written in popular terms.

G. E. Dunn and B. I. Miller, *Atlantic Hurricanes,* Louisiana State University Press, Baton Rouge, La., 1960, 326 pp.

Thorough book on all aspects of hurricanes written by two Weather Bureau authorities. Suitable for use by students and scientifically inclined laymen.

I. R. Tannehill, *Hurricanes: Their Nature and History*, Princeton University Press, Princeton, N.J., 1954, 308 pp.

A discussion in popular terms of the essential facts and theories regarding hurricanes.

G. T. Trewartha, *An Introduction to Climate*, (Third Edition), McGraw-Hill Book Co., New York, 1954, 402 pp.

A good introduction to all aspects of meteorology and climatology. Chapter 5 discusses the formation and structure of storms.

S. Petterssen, *Introduction to Meteorology*, (Second Edition), McGraw-Hill Book Co., New York, 1958, 327 pp.

A text for students who are considering meteorology as a profession, by a world authority.

H. Riehl, *Tropical Meteorology*, McGraw-Hill Book Co., New York, 1954. Chapter 11.

A scholarly discussion of all aspects of hurricane formation by a world authority on the subject. This material is for the intermediate and advanced student.

H. R. Byers, *General Meteorology*, (Third Edition), McGraw-Hill Book Co., New York, 1959.

An introductory text for serious students of meteorology, by a world authority.

Weatherwise, American Meteorological Society, 45 Beacon Street, Boston, Mass. Six issues per year.

A journal for students and laymen. It contains short articles on all aspects of the weather and frequently has reports on tornadoes, hurricanes and cyclones. Articles are written in popular terms by experts. An excellent means for keeping up-to-date on many of the latest weather advances.

The U. S. Weather Bureau prepares many short reports and brochures which are published by the Government Printing Office. They are written in popular terms. A few of the interesting titles are listed below along with the catalog numbers and prices. Rules of the Government Printing Office require remittance in advance. Unfortunately postage stamps and foreign money will not be accepted.

Storm Detection Radar, How It Helps the Pilot, Cat. No. C 30.65:6 ($.05)

Thunderstorms, Part I, Cat. No. C 30.65:7 ($.05)

Thunderstorms, Part II, Cat. No. C 30.65:8 ($.05)

The Hurricane, Cat. No. C 30.2:H 94/2/956 ($.20)
A description of hurricanes.

Hurricane Warnings, Cat. No. C 30.2:H 94/3/958 ($.05)
A brief description of hurricanes, hurricane warnings and hurricane safety precautions for Gulf and Atlantic coast areas.

It Looks Like a Tornado, Cat. No. C 30.6/2:763 ($.10)
An aid for distinguishing tornadoes from other cloud forms.

Tornadoes, Cat. No. C 30.2:T 63/4/957 ($.05)
A description of what tornadoes are and what to do about them.

INDEX

Adiabatic lapse rate (*see also* Lapse rate), 31–33
Adsorption of solar radiation by water vapor, 28
Advection, 25
Air (*see also* Turbulence), clear, 28–32; cooling by expansion, 32; deflection of, 44–46; density of, 41, 50; disturbances of, 102–3; moist, 32–33; treated as fluid, 27; vertical motions in, 23–24, 28–32, 49–52, 71–72, 90–91, 102–3, 108, 137, 143; weight of, 30, 50–51
Airplanes, damage to, 59, 73; flight over arid regions, 23, 35; flight through thunderstorms, 57–59, 73–74; observations by, 12, 62, 118
Angular momentum, defined, 44; in tornadoes, 92
Anticyclones (*see also* High pressure), 139; and Coriolis force, 47; defined, 130–31

Archimedes, 50
Arid regions, 23, 31, 35
Atlantic Ocean, 101, 111–15, 118, 138
Atom bomb, 21
Audrey (hurricane), 99–100, 120

Barographs, 78–79
Beebe, R., 90
Benard convection, 25–27, 35
Bibliography, 147–49
Bigler, S. G., 95
Bjerknes, J., 132
Black River, Jamaica, hurricane, 105–6
Blizzards, 130, 138
Braham, R. R. Jr., 22, 57, 58, 147
Brooks, E. M., 91
Bubbles in convection clouds, 60–62
Bubble theory of a thunderstorm, 60–62
Buoyancy, in cumulus clouds, 53, 56; in dust devils, 35; force of, 50–51; in hurricane develop-

Buoyancy (cont'd)
ment, 103, 109; negative, 52

Byers, H. R., 57, 58, 147, 148

Cameron (Louisiana) hurricane, 99
Caribbean Sea, 101, 112–13, 115, 118
Carr, J. A., 78
Cedar Keys (Florida) hurricane, 115
Cells, in Benard convection, 25–27, 35; formation in the atmosphere, 34–35; in thunderstorms, 57
Centrifugal force, defined, 42; formula for, 43; and wind direction, 48–49
Centripetal force, defined, 42; and kinetic energy, 44
Climatological Data (a publication), 83–84
Cloud wall, 106
Clouds, buoyancy of, 51–52, 56; convective, 32–33, 36–38, 53–54, 60–62, 71, 94, 111; cumulonimbus, 54; cumulus, 37–38, 53–54; cumulus congestus, 54; in cyclones, 136–37; droplets in, 64; electrification of, 68–71; formation of, 22, 136–37; in hurricanes, 104, 106, 110–11; lines of, 37–38; vertical speed of, 51; virtual temperature in, 51–52
Cloudstreets, 37–38
Cold fronts, 133–37
Cold weather, 138
Columnar theory of a thunderstorm, 57–60

Condensation (*see also* Heat, of condensation), into clouds, 32, 136–37; in frontal formations, 135; of water vapor, 22, 101–2
Conservation of angular momentum (*see* Angular momentum)
Convection, bubble, 60–62; cellular, 25–27, 34–35; in clear air, 28–32; in cloudstreets, 37–38; columnar, 57–60; defined, 25; "forced," 36; in hurricanes, 102; lines of, 27–28; in moist air, 32–33, 37; over mountains, 36–37, 53; patterns of, 25–28, 34–35; in thunderstorms, 53–54; in unstable air, 53–54, 89, 91; upper limit of, 32
Coriolis, G. G., 46
Coriolis force, 44; a "fictitious force," 45; formula for, 46; and geostropic wind, 48; and gradient wind, 49; and hurricanes, 47, 102; and rotation of vortices, 47
Cumulonimbus, 54
Cumulus, 37–38; growth into cumulonimbus, 53–54; "stage" of thunderstorms, 57–58
Cumulus congestus, 54
Cyclone cellars, 96
Cyclones (*see also* Hurricanes, Low Pressure), clouds in, 23, 136–37; and Coriolis force, 47; defined, 129; dimensions of, 24, 132, 137; duration of, 132; effects of, 129–30; extratropical, 116, 129, 132; forecast-

Cyclones (cont'd)
ing, 141; formation of, 132–37; height of, 132; kinetic energy in, 21; paths of, 138–41; patterns of pressure in, 139; rainfall in, 23, 129–30, 137, 138; regions of formation of, 137–41; rotation of, 129, 135; speed of, 139; tornado, 91; wind velocity in, 131

Davis, W. R., 125
Deserts (see Arid regions)
Disturbances in wind field, 102–3, 121
Downdrafts, in squall lines, 74; in thunderstorms, 50, 52, 59–60, 71; velocity of, 63
Dunn, G. E., 147
Dust devils, and Coriolis force, 47; formation of, 23, 35–37; height of, 36; kinetic energy in, 21; rotation of, 47, 81; similarity to waterspouts, 94
Dyersburg (Tennessee) tornado, 78

Echo, in radar, hook-shaped, 95
Electrification (see also Lightning), in thunderstorms, 68–71; in tornadoes, 92
Electronic computers, 12, 141, 145
Elliot, John, 107
Energy (see Kinetic energy)
Equator, magnitude of Coriolis force at, 46
Evaporation, in clouds, 52, 60, 90; in convection, 27
Extrapolation, 120
Eye (of hurricane), 100, 105–9, 110

$F = ma$, 51, 144
Floods, 74, 124, 126, 130
Flora, S. D., 85, 147
Fluids, convection in, 25–27, 60; weight of body in (Archimedes' Principle), 50
Force (see Buoyancy, Centrifugal force, Centripetal force, Coriolis force, Pressure force)
Forecasting, of cyclones, 141; of hurricanes, 120–22; of tornadoes, 89–91, 94
Formulas, for angular momentum, 44; for centrifugal force, 43; for Coriolis force, 46; $F = ma$, 51, 144; for pressure change per unit distance, 41; for pressure force, 41
Franklin, Benjamin, 67
Friction, 21, 40; defined, 41–42; effect of, 49, 50; between ice particles, 70
Fronts (see also Polar front), 50; cold, 133–37; defined, 133; formation of, 132–35; occluded, 134–36; stationary, 134; warm, 133–37
Funnels (see Tornadoes, funnels of)

Geophysical Institute (Bergen, Norway), 132
Geostropic wind, 48
Girls' names for hurricanes, 101
Gliders, 36, 38
Gracie (hurricane), 125
Gradient wind, 49
Gulf of Mexico, 89, 99, 101, 112, 118, 135

Gunn, R., 70

Gusts (*see also* Winds), 23; dimension of, 24; kinetic energy in, 21; in squall lines, 74; of surface winds, 71–72; in thunderstorms, 52, 59, 60, 71–72; velocity of, 72

Hail, against airplanes, 57, 73; and electrification in thunderstorms, 70; formation of, 64–66, 91; frequency of, 72; layers of ice in, 65–66; size of, 56, 62; speed of fall of, 56, 66

Haze, as manifestation of instability, 31–32

Heat (*see also* Convection), of condensation, 22, 23, 32, 92, 101–2, 104–5; latent, 102; sensible, 102; as source of energy, 22; transfer from ground, 23, 28, 34, 37; transfer from water, 34, 102

High pressure (*see also* Anticyclones), 24, 40

Hook-shaped echo, 95

Humidity, 23, 32, 135; effect on stability, 30

Hurricanes, clouds in, 104, 106, 110–11; and Coriolis force, 47, 102; damage by, 99–100; description of, 99–101, 105–7; dimensions of, 24, 104, 106, 108, 131; duration of, 100, 131; eye of, 100, 105–9, 110; frequency of, 101; forecasting of, 120–22; formation of, 101–4, 136; input energy in, 21; kinetic energy in, 21; locating, 116–18; movements of, 113–16; naming of, 101; observation of, 118; paths of, 111–13; precautions against, 98, 125–27; pressure in, 117, 124–25; rainfall in, 104, 105, 106, 110–11, 115, 123–24; season of, 101; speed of, 111–13; temperature in, 108–9; and thunderstorms, 110, 118, 125; tornadoes in, 125–26; waves produced by, 107, 115, 122–23, 124; weakening of, 104–5, 116, 123, 132; wind velocity in, 100, 104, 106, 114–15, 131

Hydrogen bomb, 21

Ice, in hail, 65–66; in thunderstorm clouds, 51–52, 64–66, 68–71

Ice nuclei, 64

Illinois State Water Survey, 74, 95

Indian Ocean, 100

Instability, and adiabatic lapse rate, 32–33; convective, 89, 91; and cumulus clouds, 53–54; defined, 29–30; over deserts, 31; factors leading to, 30–31, 89, 143; in frontal development, 135, 137; in hurricane development, 102–5; thermal, 27–28, 37, 89; in thunderstorm development, 53–54, 57–62, 137; in tornado development, 89–90, 93; in waterspouts, 94

Ionization in lightning, 67, 71

Isobars, 40–41, 42, 134; la-

Isobars (*cont'd*)
beling of, 48; in weather maps, 113, 140; and wind direction, 49, 114

Jet winds, 90–91
Justice, A. A., 81 n.

Keller, Will, 81–82
Kinetic energy, in atom bombs, 21; and centripetal force, 44; in cyclones, 21; defined, 20; in dust devils, 21; in gusts, 21; in hurricanes, 21; sources of, 22–23; in thunderstorms, 21, 22; in tornadoes, 21
Kuettner, J., 37

Lapse rate (*see also* Adiabatic lapse rate, Moist adiabatic rate), in arid regions, 35; in convection, 26–28; defined, 26; and instability, 27, 29, 32, 89; inversion of, 89; negative, 26; positive, 26; in tornado conditions, 89–90
Latitude, effect of Coriolis force at various degrees of, 47
Ligda, M. G. H., 72
Lightning, against airplanes, 73; cause of, 67–71; ionization in, 67, 71; return strokes of, 67; step-leaders in, 67; and thunder, 66, 68; in thunderstorms, 57, 60, 66–71; in tornadoes, 92
Line squalls (*see* Squall lines)
Lomonosov, Mikhail V., 67
Low pressure (*see also* Cyclones), 40, 50; filling of,

116; in hurricanes, 117, 124–25; in Northern Hemisphere, 48–49; in tornadoes, 78–80, 82–84
Ludlam, F. H., 60, 66

McDonald, J. E., 56
Malkus, J. S., 38
Maps, of paths of hurricanes, 112; weather, 113, 129, 140
Measurements (*see* Observations)
Mediterranean Sea, 138
Meteorology, status as a science, 11–12, 143
Miller, B. I., 147
Millibars, 48, 114
Missouri, damage from tornado, 76
Modification of weather, 22
Moist adiabatic rate (*see also* Lapse rate), defined, 32; in tornado conditions, 89
Molecular weight, 30, 51
Mountains, cloud formations over, 37, 53; convective currents over, 36–37; in formation of cyclones, 138

Nagasaki atom bomb, 21
National Hurricane Project, 118
New England hurricane (1950), 115
New Mexico Institute of Mining and Technology, 69
Norton, G., 113, 114 n.

Observations, by airplanes, 12, 62, 118; difficulty of measurement in, 12–13, 24, 62; of hurricanes,

Observations (*cont'd*) 118; lack of, 144–45; of pressure in tornadoes, 78–79; radar, 54, 56, 72, 95, 110, 118, 144; radiosonde, 90; of rainfall, 63; by rocket, 38; by satellites, 24, 38, 145

Occluded fronts, 134–36

Oceans, as birthplaces of tropical storms, 101–5; cloudstreets over, 38; as sources of convection, 28, 34; tropical, 28, 35, 38, 89; waves in, 107, 115, 122–23, 124

Pacific Ocean, 38, 100, 138
Palmen, E., 108
Petterssen, S., 148
Plow winds, 72
Polar front, defined, 135; over Pacific Ocean, 138; position of, 139; theory of cyclone formation, 132, 135–36, 139

Precautions, against hurricanes, 98, 125–27; against tornadoes, 83–84, 95–98

Precipitation (*see also* Hail, Sleet, Snow, Rain), as source of thunderstorm electrification, 68, 70

Pressure (*see* High pressure, Low pressure)

Pressure change per unit distance, 41

Pressure gradient, 41

Pressure force, 40–42; formula for, 41

Radar, 144; airborne, 73; hook-shaped echo in, 95; observation of hurricanes, 110, 118; observation of thunderstorms, 54, 56, 72; observation

of tornadoes, 95

Radiation (*see* Solar radiation)

Radiosonde observations, 90

Rain, from cyclones, 129–30, 137, 138; formation of, 63; in hurricanes, 104, 105, 106, 110–11, 115, 123–24; measurement of, 63; from squall lines, 74; in thunderstorms, 58, 60, 62–64; in tornadoes, 96

Return stroke, 67

Reynolds, G. W., 86

Riehl, H., 38, 101–2, 124, 148

Rockets, 38

Rocky Mountains, 138

Rotation (*see also* Coriolis force), of air in circle, 42–44; of cyclones, 129, 135; of dust devils, 47, 81; of earth, 44–49, 135; of tornadoes, 76, 80–81, 87–88

St. Louis tornado, 79

St. Louis University, 91

San Felipe hurricane, 123

Satellites (Tiros I), 24, 38, 145

Scottsbluff tornado, 75

Scorer, R. S., 60

Sea gulls, 34, 36, 37

Simpson, G. C., 70

Sleet, 62

Smoke trails, 35

Snow, 129–30, 137, 141

Solar radiation, 145; on mountains, 36; as source of convection, 28, 34–36

Squall lines, 72–74, 137

Stability (*see* Instability)

Stationary fronts, 134

Steering level, 121, 139

Step-leader, 67
Storm cellars, 96–97
Storms, in general, 20, 129–30
Stratosphere, 54, 132; as limit to thunderstorm growth, 55
Subcooled droplets, 64
Supercooled droplets, 64–66

Tannehill, I. R., 105, 107 n., 112, 148
Temperature (*see also* Lapse rate), 135; of cloud droplets, 64; difference between cumulus and surrounding air, 54; and dust devils, 23; in hurricanes, 108–9; and stability of air, 30, 50; in stratosphere, 55; in thunderstorms, 71; in tornadoes, 93; virtual, 51–52
Texas A. & M. College, 95
Thermal instability (*see* Instability)
Thermals, 89; in clear air, 33–37; defined, 33; over deserts, 35–36; speed of, 33
Thunder, cause of, 67–68
Thunderstorm cells, 57
Thunderstorm Project, 57, 61
Thunderstorms (*see also* Hail, Lightning, Rain), cells in, 57–60; defined, 66; diameter of, 60; downdrafts in, 33, 50, 54–56; duration of, 60; electrification of, 68–71; flight through, 57, 59, 73–74; in frontal development, 137; heights of, 54–56; gusts in, 52, 59, 60, 71–72; in hurricanes, 110, 118, 125; kinetic energy in, 21, 22; life cycle, 57–60; lines of, 72–74, 90, 110; stages of, 57–60; temperature fall in, 71; theories of formation of, 57–62; and tornadoes, 22–23, 74, 90–91; updrafts in, 33, 50, 54–56; weight of condensed water in, 22
Tiros I, 24, 38, 145
Tornado alerts, 94–95
Tornado cyclones, 91–92
Tornadoes (*see also* Waterspouts), angular momentum in, 92; damage from, 75–76, 82–85; descriptions of, 77–82, 87–88; detection of, 94–95; difficulties of observation of, 12, 24; duration of, 92; energy in, 21–22; explosive effects of, 82–85; formation of, 74, 88–93; frequency of, 72, 76–77; funnels of, 77–82, 91–92, 96; in hurricanes, 125–26; kinetic energy in, 21; lapse rates in, 89–90; lightning in, 92; noise of, 81–82, 96; precautions against, 83–84, 95–98; pressure in, 78–80, 82–84; regions of, 77, 88–89; rotation of, 76, 80–81, 88; size of, 77; speeds of, 97; and thunderstorms, 22–23, 74, 90–91; time of day of, 93; unusual effects of, 85–88; wind velocity in, 80, 84–86, 100, 131
Trewartha, G. T., 148
Trade winds, 103

Tropics (*see also* Hurricanes, Oceans, Tropical), cloudstreets in, 38; stable layers of air in, 28
Tropopause, defined, 55; height of, 56
Troposphere, 55
Turbulence, 23, 35; over arid regions, 31–32; formation of, 29; in thunderstorms, 57, 59, 60, 73
Twister (*see also* Tornadoes), 77
Typhoons (*see also* Hurricanes), 100

U. S. Air Force, 90, 118
U. S. Navy, 118
U. S. Weather Bureau, 79, 90, 118, 119, 126–27, 149
University of Chicago, 22, 57
Updrafts (*see also* Thermals), 24, 50, 51–52; and hail, 66; speeds of, 33, 54, 56; in thunderstorms, 57, 59, 61–62, 73; in waterspouts, 93–94

Van Tassel, E. V., 76 n., 87
Virtual temperature, 51–52
Vonnegut, B., 70, 92
Vortices (*see also* Anticyclones, Cyclones, Hurricanes, Thunderstorms, Tornadoes, Waterspouts), 20, 23; and angular momentum, 44; differences between types of, 131–32; energy in, 20–22; production of, 143–44; rotation of, 47; as signs of instability, 20

Warm fronts, 133–37

Warm sector, 137
Water vapor, 50; adsorption of solar radiation by, 28; condensation of, 22, 101–2; weight of, 30, 51
Waterspouts, 93–94
Waves (in air), 103, 135, 139
Waves (in ocean) (*see* Hurricanes, waves produced by)
Weather, maps, 113, 129, 140; modification of, 22
Weight, of air, 30, 50–51; of body in fluid, 50; of condensed water in thunderstorm, 22
West Palm Beach hurricane, 117
Williams, (Rev.) J. J., S.J., 105
Willy-willies (*see also* Hurricanes), 101
Wilson, C. T. R., 71
Winds (*see also* Gusts), and centrifugal force, 43; in cyclones, 131; defined, 39; disturbances in, 102–3, 121; duration of, 24; in frontal formations, 134–35; geostropic, 48; gradient, 49; in hurricanes, 100, 102–6, 111, 114–15, 131; jet, 90–91; plow, 72; in tornadoes, 80, 84–86, 131; trade, 103; velocity of, 24; why they blow, 39–40
Woodcock, A. H., 34, 37
Woods Hole Oceanographic Institution, 34
Workman, E. J., 69
Wyman, J., 35

Zoch, R. T., 117

CPSIA information can be obtained
at www.ICGtesting.com
Printed in the USA
BVOW03*0625150417
481126BV00004B/84/P